BRINGING THE FOOD ECONOMY HOME

BRINGING THE FOOD ECONOMY HOME

Local Alternatives to Global Agribusiness

Helena Norberg-Hodge
Todd Merrifield
Steven Gorelick

Bringing the Food Economy Home: Local Alternatives to Global Agribusiness

Kumarian Press, Inc.
1294 Blue Hills Avenue
Bloomfield, CT 06002 USA
(+1) 860-243-2098

Zed Books
7 Cynthia Street
London N1 9JF, UK
+44 (0) 20 7837 4014

Fernwood Publishing Co. Ltd.
P.O. Box 9409, Stn. A
Halifax, NS, B3K 5S3, Canada
(+1) 902-422-3302

Design by Nicholas A. Kosar. Copyedited by Beth Richards.
Proofread by Jody El-Assadi. Indexed by Barbara DeGennaro.
The text of this book is set in Adobe Sabon.

Printed in the United States of America on acid-free paper by
Thomson-Shore, Inc.
Text printed with vegetable oil-based ink.

∞ The paper used in this publication meets the minimum requirements
of the American National Standard for Information Sciences—Permanence of
Paper for Printed Library Materials, ANSI Z39.48–1984.

Published in the UK by Zed Books in 2002.
A CIP record for this book is available from the British Library.

UK ISBN 1 84277 233 3 pb

National Library of Canada Cataloguing in Publication Data
Norberg-Hodge, Helena.
Bringing the food economy home : local alternatives to global agribusiness

Includes bibliographical references.
ISBN 1–55266-082-6
1. Agriculture. 2. Agriculture—Economic aspects. 3. Sustainable agriculture.
4. Agriculture—Environmental aspects. 5. Farms, Size of. I. Merrifield, Todd
II. Gorelick, Steven M. III. Title.
HD9000.5.N67 2002 630 C2002-900314-8

Library of Congress Cataloging-in-Publication Data
Norberg-Hodge, Helena.
Bringing the food economy home : local alternatives to global
agribusiness / Helena Norberg-Hodge, Todd Merrifield, Steven Gorelick.
p. cm.
Includes bibliographical references and index.
ISBN 1–56549–146–7 (pbk: alk. paper) — ISBN 1–56549-147-5 (hardcover: alk. paper)
1. Food—Marketing. 2. Farmers' markets. I. Merrifield, Todd, 1969–
II. Gorelick, Steven. III. Title.
HD9000.5 .N596 2002
338.1'9—dc21 2002005518

11 10 09 08 07 06 05 04 03 02 10 9 8 7 6 5 4 3 2 1

First Printing 2002

Contents

List of Illustrations

Acknowledgments

We would like to thank Jules Pretty and Katharine Deighton of the Centre for Environment and Society at the University of Essex for helping to produce an early draft of this document. Much of the research used in this book comes from that draft.

We also gratefully acknowledge the help of Anja Lyngbaek, John Page, Becky Tarbotton, Ben Savill, Lindsay Toub, and Maya Mitchell, all of the International Society for Ecology and Culture. Helpful insights were also provided by Eve Sinton, Suzanna Jones, Brian Tokar, Miyoko Sakashita, Stephanie Roth, Karen Shaw, and Forrest Foster. A. V. Krebs's newsletter, *The Agribusiness Examiner*, provided a great deal of valuable information about US-based food corporations. Finally, we would like to honor the work of Alan Lepage in Barre, Vermont, and Aba Lagruk in Ladakh, India—two among the thousands of farmers worldwide whose local knowledge makes local food possible.

The International Society for Ecology and Culture (ISEC) is a nonprofit organization that promotes locally based alternatives to the global consumer culture. For more information, contact us at:

ISEC
Foxhole, Dartington
Devon TQ9 6EB
UK

ISEC
PO Box 9475
Berkeley, CA 94709
USA

Web site: www.isec.org.uk

1

From Local to Global

Counting all the people negatively affected by the global food system . . . we are really the majority of the people in the world.

—Peter Rosset, Executive Director, Food First

FOOD IS AT THE CENTER OF A STORM the world over. Farms in the North are going under in record numbers, even as farmers in the South are being removed from the land by the millions. Food scares occur with increasing regularity, leading many to wonder whether their meals are safe to eat. Genetically altered crops have been planted on much of America's farmland, angering consumers and environmentalists, and setting off trade disputes with Europe and Japan. Corporations are tightening their hold over the world's food supply, inciting farmers and other citizens around the world to call for boycotts, to attack fast-food chains, and to uproot genetically engineered crops.

All of this turbulence has its origins in the industrialization and globalization of food and farming. With food reduced to a commodity in a volatile market, farming is becoming ever more specialized, capital-intensive, and technology-based, and food marketing ever more globalized. These trends are proving disastrous for consumers, farmers, local economies, and the environment; nonetheless, most governments intend to accelerate the process, with policies that aim for higher exports and lower barriers to trade, more chemicals and more genetic engineering.

There is, however, an opposing current—a small but rapidly growing groundswell of support for local food systems. Consumers and farmers are forging links to promote smaller-scale,[1] more diversified, and ecologically sound agriculture. These groups favor foods grown nearby, rather than global commodities mass-produced thousands of miles away.

ISEC/Steven Gorelick

This movement is steadily growing from the grassroots, making its presence felt in the exponentially rising demand for organically produced food; in the growing popularity of farmers' markets that emphasize local varieties of fresh food; and in the eagerness with which farmers and activists in both North and South are seeking to develop more sustainable forms of agriculture.

The local food movement is all the more important because it has received almost no support from policymakers. On the contrary, governments everywhere promote the further globalization of food. They continue to pave the way—both literally and figuratively—for huge megamarkets selling food that has traveled halfway around the world. Farm agencies still promote monoculture and the use of pesticides, chemical fertilizers, high-yield hybrids, and genetically modified seeds. Schools, farm bureaus, and development agencies neglect diverse regional farming methods, promoting instead a single high-tech version of agricultural progress.

If the local food movement is spreading without help from above, in the North it is also doing so despite the fact that most consumers have very little information about the food they eat, and have almost no exposure to farming or rural life. For decades—for generations in some countries—most people in the North have been urbanized to a degree that leaves them largely isolated from nature's processes and from the countryside where their food is produced. Supermarket chains and agribusinesses deepen this separation by failing to indicate where and how a particular food was produced. Schools so rarely address the crucial subjects of food and farming that many children are left believing that food simply comes from the supermarket. The little information that does make it into school curricula is often supplied by corporate agribusinesses and other food industry interests, and is therefore woefully inadequate and shamelessly biased.

Nonetheless, awareness is steadily growing that global food is altogether too costly—to our health and that of our children, to the environment, and to local economies everywhere. Mad Cow Disease, outbreaks of food poisoning, pollution of land and water by agricultural chemicals, the decline of rural livelihoods in both North and South—these are but a few of the reasons why people are beginning to question the entire global food system and the premises on which it is based.

People are also beginning to realize that relying more on locally grown, organic foods can help solve a whole range of social and environmental problems at the same time. While enjoying the health benefits of preparing and eating fresher, more wholesome foods, they are also discovering the sheer pleasure of shopping at farmers' markets, of knowing the people who produce their food and of connecting more closely with where they live.

Shortening the links between farmers and consumers may in fact be the most strategic and enjoyable way to bring about fundamental change for the better. A world in which everyone is well fed with local, fresh foods would be a world where everyone has more power, community, and contact with nature. For many years now, colonialism and economic development have taken the world in exactly the *opposite* direction—separating not only producers from consumers, but all of us from the natural world. The question now is: Do we continue down the path of global monocultures, or do we start to shift direction?

The Global Food System

Those two options are represented by fundamentally different types of food systems. The *global* system is characterized by large-scale, highly mechanized, monocultural, and chemical-intensive methods with production oriented toward distant and increasingly global markets. The abundant use of external inputs, large machinery, and long-distance transport and communications infrastructures make this system extremely capital- and energy-intensive.

This food system is also characterized by a heavy reliance on the knowledge and technology generated by a small number of Western-style institutions. The goal is ever-increasing agricultural efficiency—defined as maximizing the yield of a narrow range of globally traded commodities, while minimizing human labor. Immense research and development efforts, many at public expense, are directed toward that end. Too often, the resulting technologies are promoted to farms irrespective of local ecological and social conditions. This has led to the reshaping of agricultural products, landscapes, and diverse cultural traditions to suit the available technologies, and the homogenizing of nature and culture to better serve the global economy. Although variants are found within the global food system, its fundamental characteristics—largely determined by technology and international market forces—are the same everywhere.

Locally Adapted Food Systems

In response to this single globalizing model, many local food initiatives are emerging around the world. These are typically oriented toward local and regional consumption, with relatively short distances—or food miles—between producers and consumers. In many cases, the two are directly linked. Having evolved within a particular social, economic, and environmental context, these new food systems in many ways mirror those of traditional cultures. In the South, in fact, one can still find thousands of indigenous, traditional, or vernacular agricultural systems—relatively small-scale and resource-conserving local food systems, each one adapted to a specific place.

Today, remnants of indigenous agriculture are found primarily in those areas of the South deemed not suitable for industrial farming. Although it is often believed that industrial agriculture feeds the world, as much as 35 percent of the world's population (about two billion people) were still directly supported by this forgotten agriculture in the mid-1990s.[2] To a much lesser extent, traditional systems also survive in more fertile lowland areas of the South, wherever farmers have resisted the dictates of overseas development experts and government agriculture extension agents.

Intentionally or not, the local food movement in the North often mimics the principles underlying these traditional food systems. In many cases, people are actively seeking the remnants of their own farming heritage and combining them with more recent advances in small-scale organic agriculture.

Centuries of Agricultural "Progress"

Over the last several hundred years, thousands of diverse, locally adapted agricultural systems around the world have been replaced by a single, globalized food system. Among the indicators of this shift are a dramatic reduction in the number of farmers and a concomitant expansion in the size of farms; a huge increase in the size and reach of agricultural markets; and a tightening control by transnational corporations over the world's food supply.

How has this happened?

In many places, the roots of these changes can be traced back at least 500 years, to the era of European conquest. As large parts of Africa, Asia, and the Americas were conquered and colonized, indigenous cultural systems were systematically swept away. Local models of food production—many of them successful at sustainably meeting people's needs—were often replaced by farms that produced food for the colonizers, or by plantations that exported food and fiber back to Europe. Farms on which families and communities raised their own food without producing a marketable surplus were of little use to the colonial powers and were converted wherever possible to more profitable uses. Most often, the best agricultural land was appropriated by the colonizers, while the task of feeding people locally was relegated to ever more marginal lands.

While European colonization doomed a great many local food systems in the South, locally adapted systems in the North did not fare much better. Sometimes the causes were similar, as when the enclosure movement in seventeenth- and eighteenth-century England privatized what had once been common land, thus eliminating the rights of villagers to farm, hunt, and gather resources as they had traditionally done.

The Industrial Revolution further undermined local food systems by drawing people away from the rural sources of their food, into urban areas where their food had to be imported from the countryside, and increasingly, from abroad. New agricultural technologies, meanwhile, steadily reduced the number of jobs available in local food systems and sapped the economic vitality of rural areas. From the earliest factory-made farm equipment to today's mammoth grain combines and tomato harvesters, these technologies have invariably been designed to reduce investors' and industrialists' expenditure on human labor, rather than to improve the well-being of farm laborers and their communities.

Farmers found that an increasing portion of their income went to equipment manufacturers, and eventually to fossil fuel producers. A fully-mechanized farmer may have been able to bring as much to market with less help than before, but that amount was no longer enough: ever-rising equipment and input costs meant that far more production was needed just to break even.

Like the craft workers displaced by the new industrial factories, farmers as a whole were systematically replaced by machines and were consequently forced into the emerging urban slums. The result can be seen in the falling

USDA/Dave Hein

proportion of England's population engaged in agriculture: in the eighteenth century some 40 percent were on the land; by 1900 this number had fallen to 8 percent, and today it is only 2.5 percent.[3]

The new technologies led to other changes on the farm. To take full advantage of their equipment, farmers were impelled to plant larger and larger expanses of machine-friendly monocultures and to homogenize their farmland by cutting down trees, ripping up hedgerows, bulldozing rock outcroppings, and ignoring the specific characteristics of each field. In other words, farms were shaped to fit the technology.

These trends were driven by more than just mechanical technologies. Other developments, such as hybrid seeds, chemical fertilizers, herbicides and pesticides[4] had a similar impact. Farmers who traditionally saved seed from one year's crop to plant the next, who used farm-produced manure to maintain the fertility of their fields, and who relied on companion planting, rotations, and biological controls to limit weeds and pest damage, instead began using increasing amounts of purchased inputs. Not only did this shift lead to a long-term deterioration in the quality of farmland and an all-too-familiar poisoning of the environment, it also forced farmers to continually increase their production to pay for inputs they previously did not need.

Since conventional economic pricing does not include the huge hidden or externalized costs of industrial food production—from topsoil loss and air pollution to pesticide-induced cancer—technology-driven development

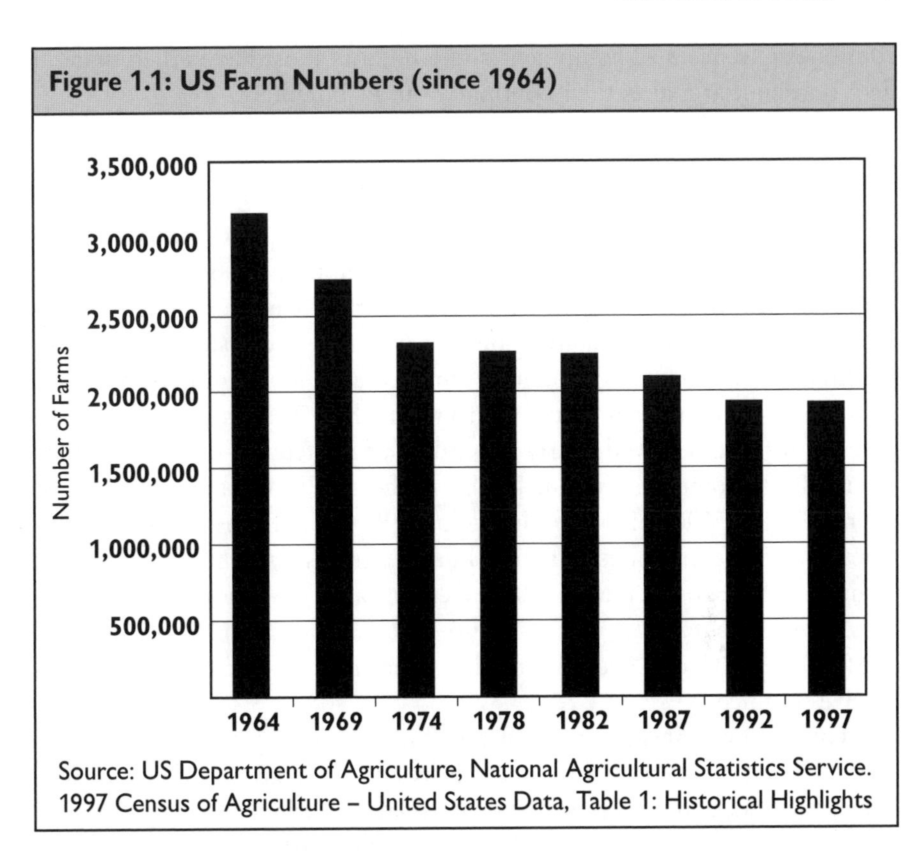

Source: US Department of Agriculture, National Agricultural Statistics Service. 1997 Census of Agriculture – United States Data, Table 1: Historical Highlights

led to a steady decline in the prices farmers received for what they produced. Any farmer who failed to adopt the latest technology, and whose marketable yields did not therefore rise, was still faced with the same drop in prices and was unlikely to be able to continue farming for long. Farmers who *did* adapt to each successive wave of innovations usually had to take out bank loans to pay for the new technologies, and they were continually pressed to increase their production, expand their landholdings, and seek ever more distant markets for their expanded output, simply to pay back the loans.

Today, only a relative handful of farmers in the industrialized world has survived this technological treadmill. The six founding countries of Europe's Common Agricultural Policy (CAP), for example, had 22 million farmers in 1957; today that number has fallen to about seven million.[5] Canada lost almost three-quarters of its farm population between 1941 and 1996, and the numbers are still declining.[6] In the United States, some 6.8 million farms were in operation in 1935; by 1964 the number of farms had been cut by more than half. The number is less than 1.9 million today.[7] (see figure 1.1). Among the family farmers remaining today, few have the capital to finance continual purchases of equipment and land, and most have fallen under a debt burden that makes their continued survival unlikely.

Big Farms Get Bigger

When small farms fail, their land and markets are quickly gobbled up by the largest farms, and increasingly, by corporate farms. These huge industrial-style operations are thriving not because they are more productive (they are, in fact, far less productive), but because they are systematically supported by government policies. In the United States, for example, direct farm subsidies rise with the amount of land a farmer has. As a result the largest 10 percent of farms received nearly two-thirds of the federal subsidies handed out in 2000—some $17 billion. The list of those subsidized includes at least twenty Fortune 500 companies, including IBP, Archer Daniels Midland, and Chevron.[8] The situation is similar in Europe: for instance, 80 percent of UK farm subsidies are given to the 20 percent of farmers with the largest holdings. One such farmer, whose 2,000 acres in Cambridgeshire net him £180,000 in subsidies every year, admits that it is "idiocy" that he should receive so much.[9]

These heavy subsidies make it possible for the largest farms to turn a profit, even if they sell their output at prices *below* their cost of production. When they do, commodity prices drop to levels that make survival all but impossible for small farmers who sell on global markets. Since their farms are small, the subsidies they receive are paltry; since farm commodity prices are below the cost of production, these farmers lose money year after year.

Graham Harvey, author of *The Killing of the Countryside*, argues that this is one of the great ironies of the way agriculture is now subsidized: "It doesn't merely strip wildlife from the land, it drives out the people, too. The billions of pounds in EU hand-outs, intended to aid small farms, are instead feeding a rampant agribusiness that is rapidly swallowing them up."[10]

Peter Rosset of Food First would agree. Farm subsidies in America, he says, are "basically a transfer of money from the pockets of US taxpayers to large corporate farmers."[11] As we will see in the coming pages, this is but one of the many ways government policies systematically support the global food system, to the detriment of smaller and more localized food systems.

Agribusiness Takes Over

While the steady shift toward full dependence on the global food system has undermined family farmers and impoverished their communities, the process has been a boon for corporate farms and for the agribusinesses on which farmers increasingly depend. Those agribusinesses market everything industrial farmers now need: farm equipment, fossil fuels, seeds, antibiotics, fertilizers, pesticides, and more.

Farmers linked to the global system rely on agribusinesses in another way as well. When farmers produce a small but diverse range of crops, their entire harvest can be marketed within their own local economy; but when swallowed up into the global food system, farmers need to market single crops in amounts far larger than the local economy can absorb. Those farmers can no longer market their own production and have come to depend on agribusinesses to do it for them.

Corporate agribusinesses have thus taken control of the entire food system. Not only do they supply almost all the needs of industrial farmers (in some cases owning the farms themselves), they also act as middlemen, processors, distributors, and retailers—buying, packaging, and selling food on markets that have grown to encompass the entire planet.

In countries where agricultural progress has proceeded furthest, many agribusinesses have become huge, vertically integrated enterprises, with subsidiaries profiting from every aspect of a farm's operation. For example, consider this hypothetical American farmer growing wheat and raising some cows and chickens, described by Joel Dyer in *Harvest of Rage*. The farmer purchases a new tractor from a company owned by the Cargill corporation and some irrigation equipment from a second Cargill subsidiary. He also needs seeds, chemical fertilizers, and feed for his livestock, all of which are purchased from still other Cargill subsidiaries. At harvest time he brings his wheat to Cargill's milling operation; unhappy with the price he is offered, he decides to store his crop in a grain elevator, also owned by Cargill, hoping for a price rise in the future. The storage charges are eating into any return he will eventually receive, so he sells his crop to a trading company, owned by Cargill, which ships it to Europe or Japan. His cattle, meanwhile, are sold to a feedlot owned by a Cargill subsidiary, which in turn sends them to a Cargill-owned meatpacking plant. He sells his chickens to one of Cargill's poultry-processing plants. Unfortunately for him, the prices he has received for his wheat, his cattle, and his chickens are too low for him to make ends meet, so he goes to a local bank for a loan. The bank, as it turns out, is also owned by Cargill.[12]

Although the farmer just described is imaginary, Cargill is involved in all aspects of farm production mentioned—and many more besides (see box, page 90). And while this farmer could have purchased some of his needs from a Cargill competitor, chances are this company also would have been a huge agribusiness corporation.

Because of corporate control over virtually all their needs, industrial farmers are in a precarious position: they have no leverage over their costs for inputs and equipment, nor over the price they receive for their production. As a predictable consequence, they are being tightly squeezed by profit-driven agribusinesses, whose share of the price of food has steadily risen at the expense of farmers. By 1990, only nine cents of every dollar spent on domestically produced food in the United States went to the farmer, while middlemen, marketers, and input suppliers took the rest.[13] In the decade since then, the squeeze has tightened even more: between 1990 and 1999, farmers' costs increased by 17 percent, while farmgate prices dropped 9 percent.[14] It is not surprising that so many farmers fail every year.

In a sense, farmers hooked to the global food system have become little more than serfs in a corporate feudal system. This metaphor is nowhere more appropriate than in the US hog and poultry industries (see box, page 11). According to Joel Dyer, "it's all but impossible for a farmer to raise hogs or poultry without first getting a contract with one of the monopolies that guarantee his animals will be purchased when the time comes. That's because . . . in most regions of the country there's only one buyer."

Dyer goes on to quote an independent hog farmer from Missouri, who says, "It used to be within 10 miles, you could go to five or six places every week to sell your hogs. Now you have to take them 50 miles to one place, one day of the month and take what the one corporate buyer will give you."[15]

Needless to say, corporate buyers aren't paying much: for example, poultry farmers caught in the industrial food system earn only *five cents* per bird, after expenses.[16]

Globalization

The shift from small-scale farming for local markets to large-scale industrial farming for national and international markets has been underway for centuries, in some places happening slowly and steadily, in others suddenly and rapidly. But in the last fifty years the overall process has accelerated dramatically, thanks to policies promoting economic globalization.

The outline of today's globalized economy was designed at the 1944 Bretton Woods conference, where Western leaders met to design a new financial architecture for the post-war period. The aim was to draw the world's smaller economies into the orbit of the industrial superpowers, thereby greatly increasing the number of Western-style consumers, expanding the market for manufactured goods, and assuring unfettered access to the planet's natural resources. With these ends in mind, three supranational institutions were established: the World Bank, the International Monetary Fund (IMF) and the General Agreement on Tariffs and Trade (GATT).

Corporate Feudalism, by Joel Dyer

In order to get a contract with Continental, Tyson, or ConAgra, the corporations that now control the [US hog and poultry] industry, farmers must agree to the companies' terms. And those terms are the equivalent of the farmer becoming a hired hand on his own land. . . . [The] companies tell the farmers what type of chicken houses or hog buildings they must build, forcing them into more and more bank debt as they struggle to keep up with the technological advances demanded by the company. The hog industry is particularly expensive. Modern operations consist of giant buildings, where hogs live out their entire existence without ever having been outdoors or having breathed fresh air. High-tech systems feed the animals and remove their excrement. The hog simply stands in one place and grows until it is killed . . . .

Many hog and poultry farmers no longer own any animals. The farmers get the chicks and hogs from the multinationals. Even the grain the animals are fed is provided by the company. At the end of the season, the full-grown animals are trucked to the company's processing plants where they're weighed. After rating each farmer's performance in pounds, the company deducts its charges for the chicks or hogs, feed, transportation, and any other services or products it supplied, such as propane to heat the buildings. If there's anything

USDA/Ken Hammond

left over, the farmer is compensated. The only things that the company allows the farmer to own are the heavily indebted buildings and land where the company raises "its" animals. And those buildings never get paid off. Farmers say that nearly every year, the company requires innovations in exchange for the all-powerful contract it knows the farmer must have to avoid bankruptcy. Once in the system, there's no way out except foreclosure—a sad reality that the companies exploit to the fullest.[a]

The Bretton Woods scheme called for the World Bank to provide funding for major development projects—first to rebuild war-torn Europe, then to build up the so-called underdeveloped parts of the world. World Bank projects have generally focused on the infrastructure requirements for a global trading system, including huge centralized energy plants, long-distance transport networks, and high-speed communications systems. The IMF, meanwhile, has worked to impose a standardized economic architecture—with unbridled economic growth as the foundation—on every national economy. Strict structural adjustment policies have been imposed on borrowing countries that did not adhere to that plan. At the same time, the GATT has served to increase every nation's dependence on long-distance trade by keeping tariffs low, removing other perceived barriers to trade, and turning more realms of life into globally tradable commodities.

For the South, this framework ushered in the era of development. Though ostensibly more noble in its aims than colonialism, its goals and policies actually joined seamlessly to those of the previous era, and the results have been strikingly similar: a Northern economic model, based on industrial production, trade, and economic growth, has been systematically imposed throughout the Third World. And just as in the colonial period, resources and output have steadily flowed from South to North.

One measure of the impact of these institutions is the degree to which national economies are now dependent on international trade. Since the 1940s, world trade has grown twelve-fold—almost two-and-a-half times faster than the growth in output. Imports and exports now make up a much larger proportion of economic activity than ever before, with international trade amounting to some $5.5 trillion annually.[17] Not all of this trade consists of consumer goods, military hardware, and natural resources: trade in food, too, is steadily expanding, with more than 600 million metric tons of

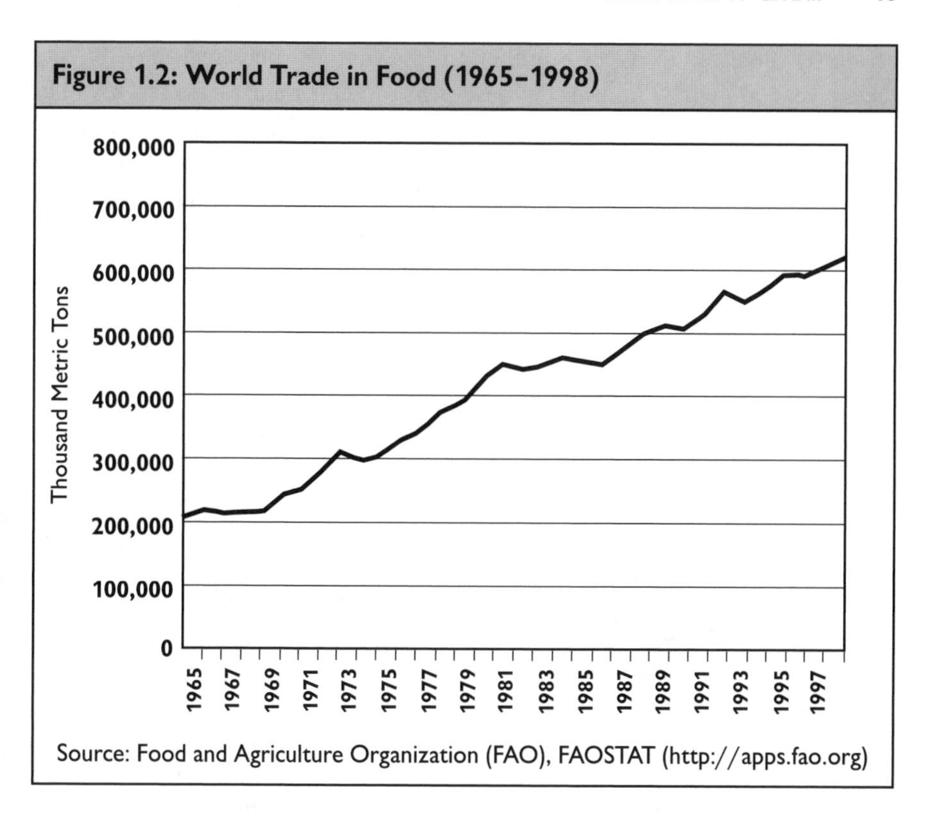

food crossing borders in 1998, triple the amount traded in 1965 (see figure 1.2).

This explosive increase in global trade has fed the growth of the trading bodies—transnational corporations (TNCs)—while systematically stripping power from local authorities. Many TNCs, in fact, are now more powerful than entire nations: a study by the Canadian Centre for Policy Alternatives showed that 51 of the 100 largest economies in the world in 1999 were corporations, not countries.[18]

While global trade and TNCs have been nurtured and supported, local economies and the food systems embedded in them have been seen as little more than anachronistic impediments to economic progress, and have been systematically dismantled. In the South, farmers have been encouraged, pushed, and cajoled by development experts into using chemicals, hybrid seeds, and farm machinery, and to orienting an ever larger proportion of their production toward national and global markets. Since these modernized farms need far less labor than traditional farms, people have been displaced from the land in phenomenal numbers. In 1979, for example, 92 percent of China's population was on the land; China's abandonment of collectivized agriculture and its efforts to integrate rapidly into the global economy have reduced that number to less than 40 percent today. In one recent year alone, 10 million Chinese peasants left their farms.[19] This may sound like an extreme example of social engineering, but it is not so different from what has happened in Northern, market-based economies: in the United States, for example, some 25 million rural inhabitants have been similarly uprooted since the end of World War II.[20]

Classical economists see all this dislocation as desirable, because it guarantees cheap labor for new urban industries. But while big businesses reap the benefits of this uprooted labor force, the inevitable costs—from poverty and unemployment in hollowed-out rural communities to the often unmanageable population explosion in the cities—must be absorbed by society as a whole.

Speeding Up the Treadmill

Free trade policies and an emphasis on exports have made farmers highly vulnerable to currency fluctuations, to recessions thousands of miles away, and to other economic forces far beyond their control. In the UK, for example, export-dependent farmers saw their incomes drop by 60 percent between 1995 and 2000, in part because of the rising value of the British pound.[21] They also saw millions of their farm animals slaughtered, as the UK government attempted to eradicate an outbreak of foot and mouth disease—and thereby salvage the market for British meat exports.[22] In the United States, meanwhile, nearly one billion bushels of grain—half the nation's harvest—found no market in 1999, largely because an economic crisis in Asia dampened demand for US products.[23] Thousands of American farmers lost their farms, thanks to farm policies that linked their fate to speculative global markets.

Those same policies also force farmers to compete with farmers further and further afield, in places where cheaper labor makes production far less costly. Farmers are thus pressured to become still more efficient by adopting newer technologies, increasing the size of their farms, and specializing in the one or two crops demanded by short-term economic trends. In other words, globalization has significantly cranked up the speed of the agricultural treadmill.

If economic globalization is allowed to continue, the treadmill will only go faster, not only for farmers but for all of us. Over the years, the GATT in particular extended its reach into new areas, steadily expanding the rights of international investors and global corporations, while drastically limiting the ability of governments to serve the interests of their citizens.

In 1994, member nations of the GATT created a new and powerful gov-

© Nick Cobbing/David Hoffman Photo Library

erning body, the World Trade Organization (WTO), to set trade rules and settle disputes. Member countries that join the WTO implicitly agree to reorient their national economies toward international trade and investment.[24] This reorganization includes the privatization of industry and the dismantling of any social program, or labor, environmental, or health regulation that could be interpreted as a trade barrier. A member country's failure to do so could expose them to sanctions and fines. The unelected bureaucrats that run the WTO thus have the power to overturn democratically enacted national and local laws, in a secret deliberative process that excludes input from farmers, consumers, labor groups, or environmental organizations. The scope of the WTO's power is alarming, to say the least. According to one estimate, as much as 80 percent of US environmental law could be challenged under WTO rules.[25]

Fortunately, people around the world are being alerted to the implications of this corporate power grab. Opposition to the WTO came to a head at the end of 1999, when trade ministers met in Seattle, Washington, to negotiate the agenda for a new Millennium Round of the GATT. The Seattle talks collapsed, partly because of massive protests outside the negotiating halls, and partly because of the unwillingness of Southern delegates inside the halls to be excluded from the decision-making process. Coming on the heels of the NGO-led derailing of the Multilateral Agreement on Investments (MAI)—which would have expanded the range of businesses in which

TNCs would be given free rein—the Seattle protests thoroughly demolished the myth that globalization is inevitable.

Since then, there have been similar protests in Washington, DC, Prague, Quebec City, Genoa—almost everywhere policymakers have met to promote the corporate agenda. By linking consumer, labor, and environmental concerns and by joining the interests of the Third World with those of communities in the North, the protesters have come to represent a large portion of the world's people. And virtually all the protestors are demanding a fundamental shift in direction.

For food in particular, such a shift is clearly needed. Centuries of agricultural progress in both North and South have taken away farmers' livelihoods, sapped the economic vitality of rural areas, and deeply damaged the environment—all while reducing the quality of our food. Although the growth of the local food movement inspires hope that fundamental change is coming, its future remains in doubt as long as government policy remains so firmly tilted against it. If government policies instead served to level the playing field, local food systems could once again supply the majority of people's food needs everywhere, just as they did not so long ago.

2

The Ecology of Food Marketing

Overflowing landfills, befouled skies, eroded soils, polluted rivers, acidic rain, and radioactive wastes suggest ample attainments for admission into some intergalactic school for learning-disabled species.

—David Orr, *Earth In Mind*

A KEY FEATURE OF LOCAL FOOD SYSTEMS is that food miles—the distances food travels before reaching the consumer—are relatively low. This means that local foods use far less energy, and produce less pollution and greenhouse gases, than food from the global system. This, in fact, may be one of the strongest arguments in favor of a shift toward local foods.

It is no longer in doubt that greenhouse gas emissions are altering global climate. Despite the pie-in-the-sky predictions of some, global warming does not mean that Scandinavia and New England will become suited to growing bananas and citrus fruits. Rather than gradual warming, weather everywhere is likely to become more unstable, unpredictable, and extreme. All over the world, in fact, people are noticing that weather patterns are changing—and not for the better. In western New York, for example, parching drought from 1997 to 1999 was followed by torrential rains in 2000, the worst in 50 years. The extreme weather "could be the final straw for many farmers," according to the state's governor.[1]

Climate change of that sort entails risks so high that it is irrational to continue business as usual, especially when that includes encouraging people everywhere to depend on food transported thousands of miles instead of food produced next door. It is a notion that borders on lunacy, yet this is exactly what government policies in almost every country promote.

As a consequence of those policies, food miles within the global food system are huge. A typical plate of food in the United States, for example, has accumulated some 1,500 miles from source to table.[2] America is not particularly exceptional in this regard among industrialized countries: a study in Germany revealed that the ingredients in a single container of yogurt had

Figure 2.1: UK Trade in Fresh Milk (1965–1998)

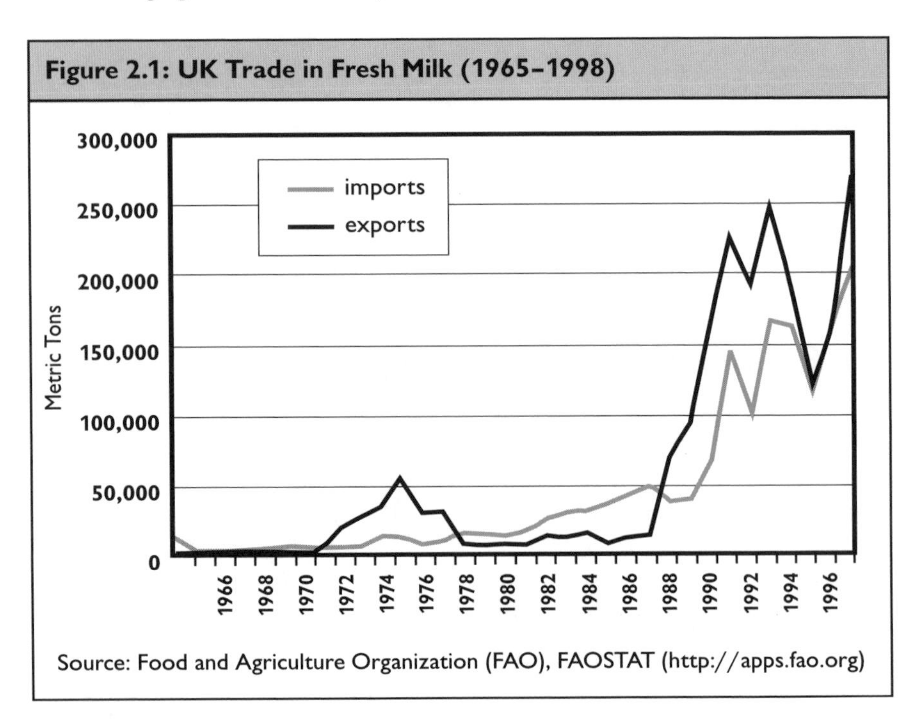

Source: Food and Agriculture Organization (FAO), FAOSTAT (http://apps.fao.org)

come from four different countries and required 1,000 km of transport.[3] Adding the transport involved in food packaging—which incorporates materials originating in distant forests and oil fields—would increase those numbers substantially.

The US Department of Transportation estimates that food and agricultural products account for 566 *billion* ton-miles of transport within America's borders each year—a sum which does not even include food shipped to or from the country.[4] Thanks to globalization, numbers like these are on the increase worldwide: in the United Kingdom, for example, the average distance that food traveled grew by more than 50 percent between 1978 and 1998.[5]

It is commonly assumed that all this food transport simply enables people to consume fruits, vegetables, and other foods unavailable from nearby sources. But the issue is not about people's right to consume food that cannot be grown locally; the fact is that a huge portion of trade in food today is unnecessary, with countries importing vast quantities of food that they themselves produce in abundance. In 1996, for instance, Britain imported more than 114,000 metric tons of milk. Was this because British dairy farmers did not produce enough milk for the nation's consumers? No, since the United Kingdom *exported* almost the same amount of milk that year, 119,000 tons.[6] (see figure 2.1).

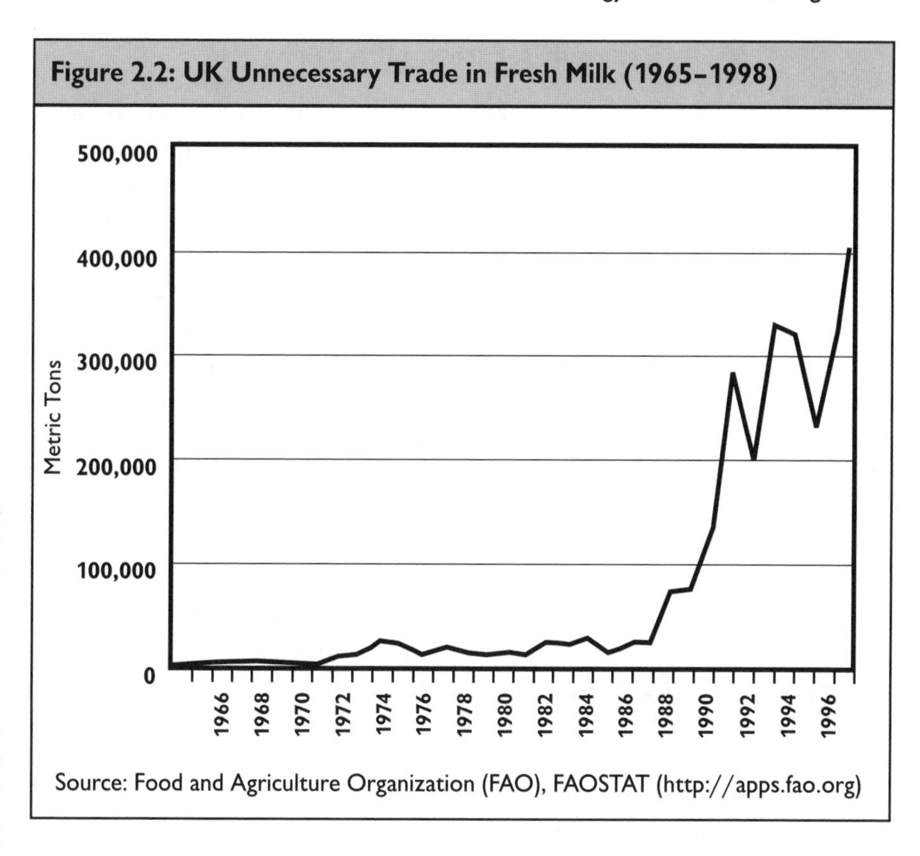

Similar numbers can be found for other foods and for other countries. Even worse, the trend toward simultaneously importing and exporting identical commodities is on the rise: in 1998, for example, the United Kingdom needlessly transported five times more milk than in 1990 and over thirty times more than in 1980 (see figure 2.2). For the most part, this excessive transport benefits only a few large-scale agribusinesses and speculators, which take advantage of government subsidies, exchange rate swings, and price differentials to shift foods from country to country in search of the highest profits. The ecological costs of these practices are too high to ignore, particularly in an era of human-induced climate change (see box, page 31).

Although proponents of free trade argue that fleets of cargo ships, trucks, and planes carrying the same commodities in opposite directions somehow leads to economic efficiency, the current system is, by any reasonable measure, absurdly inefficient. As economist Herman Daly has facetiously pointed out: "Americans import Danish sugar cookies, and Danes import American sugar cookies. Exchanging recipes would surely be more efficient."[7]

Reduced vs. Excessive Packaging

Local foods have other environmental advantages over industrial foods. Since local foods are more often consumed fresh, they usually require far less packaging, processing, and refrigeration: fresh peas, for example, require only 40 percent of the energy expended for a frozen carton of peas, and only 25 percent of an aluminum can of peas.[8]

There is also the problem of what to do with the waste that results from all the packaging required by foods transported thousands of miles. In the United Kingdom, at least a quarter of household waste is made up of packaging, two-thirds of which is used on food.[9] More and more land must be devoted to burying this huge amount of waste, because it is produced on a scale that natural processes cannot possibly absorb. Much of the packaging is non-biodegradable plastic, and even paper cannot break down in dense, poorly aerated landfills.[10] Burning all this refuse is an even worse option: trash incinerators contaminate the air with hundreds of pollutants, including carcinogenic substances such as dioxin, while leaving behind an ash residue contaminated with heavy metals and other toxins.[11]

Large-scale processors and marketers receive significant benefits from packaging: it not only protects food from the rigors of long-distance transport, but when plastered with heavily advertised brand names it can also increase the profit margins of agribusinesses and supermarkets. Since the disposal cost of the resulting waste is not paid by those businesses, they have little incentive to reduce the amount they use.[12]

Small, Decentralized Shops vs. Huge Megamarkets

The marketing of local foods is highly decentralized, which means that not only are food miles low, the distances people need to travel to purchase food are relatively low as well—often within walking distance. In some cases, farmers sell directly to consumers via box schemes and other community

Ideas That Work: Farmers' Markets

Farmers' markets were once very common throughout the world, but they declined along with the number of diversified small-scale farmers. Now they are on the rise again. According to the USDA, there were over 2,800 farmers' markets in the United States in 2000,

ISEC/Steven Gorelick

a 63 percent increase since 1994.[a] The United Kingdom went from having no farmers' markets at all in the mid-1990s, to having more than 270 at the end of the decade.[b]

Farmers' markets are sometimes held in a permanent structure but more often are on a segment of street that has been blocked to traffic. Most of the producers that sell at these markets are relatively small and tend to use fewer pesticides than do large-scale monocultures, even if they are not certified organic. Consumers have a range of in-season fresh produce to choose from, often harvested that same day. At most farmers' markets, consumers can also choose from a variety of other products, such as honey, jams and preserves, ciders and juices, bread, cut flowers, and potted plants. Many farmers' markets also have vendors selling meat and dairy products, most of which are free of the hormones and antibiotics found in products from large factory farms. Some farmers' markets even feature live music, played by either paid performers or street musicians.

These markets are community affairs, not only because of their friendly and relaxed ambience, but because the money spent there supports local enterprises and remains circulating within the community. Even though some nearby food merchants initially view the market as a potential competitor, they usually find that increased foot traffic from the market actually helps them as well. A representative of the Connecticut Department of Agriculture concurs, pointing out that a farmers' market "never, ever impacts other businesses . . . If anything, it increases their business."[c]

supported agriculture (CSA) plans, farm stands and farmers' markets, or by sales made at the farm itself. The number of farmers already selling directly to consumers varies from country to country. About 5 percent of UK farmers, 15 percent of German and US farmers, and 25 percent of French and Japanese farmers now sell directly to consumers.[13] In the United States, 19,000 farmers rely solely on farmers' markets to sell their produce.[14]

Compared with local food systems, the marketing of foods produced as global commodities is highly centralized, with the typical retail outlet being a giant supermarket or hypermarket serving a large area. These supermarkets have tremendous ecological drawbacks. To reach them, consumers often must travel significant distances—usually by car—adding to the greenhouse gas and air pollution toll. Their initial construction, including their acres of free parking, often goes hand in hand with the expansion of roads and usually entails the destruction of greenfield sites outside of established town centers.

For the sake of efficiency, huge warehouses known as regional distribution centers are increasingly used to store goods from suppliers before they are sent to retail outlets. While this form of distribution lowers costs for supermarket chains and wholesalers, it actually leads to an increase in transport, as goods must first travel to the distribution center, rather than being taken directly to the retail outlets. Furthermore, the trend toward just-in-time distribution means that each delivery contains 30 to 40 percent less produce: more deliveries are required, more energy is used, and more pollution is emitted.[15]

The growing dependence on supermarkets also raises the number of shopping trips consumers make and increases the distance each trip represents, thus increasing the likelihood that these trips will be made by car. The UK Department of Transport found that the average distance traveled per shopping trip increased by 31 percent between 1986 and 1996, while the total distance each person travels annually for shopping rose 41 percent.[16] Most of this increase was made up of car travel, which now accounts for nearly two-thirds of all transport for shopping.[17] The increased use of cars by supermarket shoppers is leading to heavy costs for society as a whole: it is estimated that the total cost of car use for an out-of-town supermarket—including air pollution, carbon dioxide emissions, noise, and accidents—is £25,000 per week higher than for an equivalent market in the town center.[18] Some studies indicate that including the cost of congestion would at least double this figure.[19]

In this regard, trends in the United Kingdom (and many other countries)

Ideas that Work
Community Supported Agriculture (CSA)

CSAs are the next best thing to growing your own food. The basic principle is simple: consumers pay the farmer in advance and receive a certain share of the produce in return. There are several different CSA models:

- In subscription farming, consumers buy a share of the harvest at the beginning of the season. A budget for the farm is drawn up and the value of each share is estimated. In this way, consumers share in the risks of farming, but in return are allowed some involvement in farm policy and some influence over what produce they receive.
- Community farms work on the same principle as subscription farms, but the farm is owned by everyone in the community, with farmers as equal partners.
- The most popular type of CSA is undoubtedly the box scheme. Here, consumers pay for a certain number of weekly food deliveries; the boxes of produce can be picked up at the farm or a central pick-up point, or they can be delivered directly to the consumers' homes. The content of each box is standard, rather than made-to-order, and is determined by what produce is ready for harvest in any particular week.

CSAs encourage agricultural diversity. Because farmers are growing for people rather than for an abstract market, they tend to grow a greater variety of produce. Most offer a mix of eight to twelve vegetables, fruits, and herbs per week, while some include value-added products such as cheese, honey, and bread.

Farmers benefit from CSAs by having an assured and stable market for their produce and by receiving payment in advance rather than when the harvest is in. Benefits to CSA members include a lower overall price for a season's worth of food and the important knowledge of where and how their food was grown.

CSAs not only enable consumers to establish a personal relationship with the farmer who grows their food, they often provide an opportunity for urban consumers to reconnect with the land. With many CSAs, the farmer welcomes visits from members, perhaps even organizing workdays that enable them to help out on the farm. Most important, consumers are able to eat produce that is much fresher and more nutritious than nearly anything they could buy at the local supermarket.

ISEC/Steven Gorelick

There are currently some 1,000 CSA operations in North America.[d] In the United Kingdom, there are now nearly 200 produce box schemes, providing food to more than 45,000 households and generating sales of £22 million.[e]

A recent study of eighty-three CSAs in the United States has shown what people value most from them. More than 60 percent of the farmers said that the most successful aspect of their operations was the strengthened bonds between people, resulting in networks that "reconnected people with the land and reconnected farmers with the people who eat the food that they grow."[f]

parallel the experience of post–World War II America, where hidden subsidies for gasoline and vast expenditures on roads and highways led to the growth of suburbs, strip developments, shopping malls, and megastores. Because of those policies, a trip to the market for a loaf of bread that once entailed a ten-minute walk now requires a ten-minute drive.

www.arttoday.com

Transport Infrastructures

An accurate reckoning of the industrial food system's ecological bill should include the costs of all the infrastructures that support it. Most of those infrastructures have been built to facilitate large-scale production and distribution of goods in general—not just food—and to support the integration of diverse local economies into a single global economy. For the most part, these infrastructures are paid for by governments, using taxpayers' money, and selectively benefit the largest, most globalized enterprises. They thus constitute a huge hidden subsidy to big business in the global economy.

The money currently spent on long-distance ground transport alone offers an idea of how heavily subsidized the global food economy is. In the United States, where there are about 3.9 million miles of public roads, another $175 billion has been earmarked for ground transportation in the next few years, with the goal of "improving access to markets worldwide."[20] The European Community, not to be outdone, is planning to spend $465 to $580 billion on a Trans-European Network that includes new high-speed rail links in and between France, England, Italy, Austria, Germany, and Spain; motorways in Greece, Bulgaria, Portugal, Spain, Ireland, Great Britain, and all the Scandinavian countries; and surface crossings between Denmark and Sweden, and between Britain and Ireland.[21] Throughout the South, scarce resources are being similarly spent. The World Bank, for example, is lending $400 million to China for highways that "will improve long-distance travel and promote trade."[22]

Beyond the profligate use of fossil fuels that they encourage, these constantly expanding transport infrastructures exact a heavy ecological toll. Modern highway construction, for example, entails the felling of forests, the filling-in of valleys, the leveling of hills, and the burying of miles of once-living ecosystem under concrete and asphalt. Large, multi-lane highways fragment the landscape as well, thereby disrupting wildlife movements and migration patterns, and limiting plant seed dispersal.

The building or expansion of a highway makes exploitation of a region far easier and can thus be an ecological death-knell for a wide swath of territory beyond the pavement. In Brazil, for instance, the trans-Amazonian highway has led to the devastation of thousands of square miles of rainforest, with gold prospectors, land speculators, cattle ranchers, and peasants dispossessed of land elsewhere pouring into regions that had previously been all but inaccessible.

The increasing numbers of ever-larger airports, shipping terminals, railway facilities, and other transport infrastructures have similar impacts. Five

Ideas That Work
Japanese Consumer Co-ops

An extraordinary movement—driven mainly by women—has been successfully linking farmers directly with consumers all over Japan. Several types of consumer cooperative have been created, including *sanchoku* groups ("direct from the place of production") and *teikei* schemes ("tie-up" or "mutual compromise" between consumers and producers).[g] There are now between 800 and 1,000 such groups in Japan, with a total membership of 11 million people and an annual turnover of more than $15 billion. These consumer-producer groups are based on relations of trust and put a high value on face-to-face contact. Some of these have had a remarkable effect on farming as well as on other environmental concerns.

The largest and best-known consumer group is the Seikatsu Club, which received the Right Livelihood Award (known as the Alternative Nobel Prize) in 1989. This group has a membership of more than 210,000 households organized into 26,000 *hans*, or local branches, all over Japan. It was set up in 1965 by Tokyo housewives seeking a way to avoid the high price of milk by banding together to buy directly from farmers. Over the next few years, they also began to purchase a wide range of pesticide-free foods, as well as clothes and cosmetics. The Seikatsu Club is now a 40 billion yen ($320-350 million) operation and employs 905 full-time staff.

Other groups are small ventures in which ten to thirty households link with a farmer who supplies food of a particular quality, usually organic. One larger group is the Young Leaves cooperative, begun by a Tokyo farmer. It now has 400 household members linked to eleven organic farmers, who supply vegetables, rice, root crops, and fruit. Members buy about 75 percent of their food through Young Leaves.

One researcher described the key components of *sanchoku* groups this way:

> Sanchoku schemes spring from moral commitments as well as commercialism, but their greatest strength lies in promoting the link between farmers and consumers. Farm walks, demonstrations and harvest festivals are organized, and weekly newsletters contain stories from the farm.[h]

Overall, the goals of many of these groups are far broader than merely obtaining high quality products at a reasonable price. The mission of the Seikatsu Club, for instance, includes seeking "to empower each and every member with a voice and role in participatory politics."[i] That part of their mission is working as well: to date, close to forty Seikatsu members have entered local politics.

ISEC/John Page

South American governments, for example, are in the process of widening, deepening, and straightening 2,100 miles of river to accommodate convoys of barges carrying soybeans and other global commodities. Known as the Hidrovia Paraguay-Parana, the project will require dredging the equivalent of four million truckloads of material from sensitive ecological areas and will threaten the Pantanal, the world's largest wetland.[23]

Energy Infrastructures

In the less-industrialized world, the energy requirements of local food systems are relatively low and can often be met entirely from nearby renewable sources: water power can run small grain mills, for example, and solar energy can be used in crop dryers. The global food system, on the other hand, has a huge appetite for energy and depends on large-scale centralized energy infrastructures to satisfy it. In addition to transport, energy is needed to process foods, to refrigerate them for their long-distance journeys, to produce the packaging in which processed foods are sold, and to power the factories in which industrial agriculture's many off-farm inputs are manufactured.

These centralized energy installations have many direct ecological costs: large-scale hydro-electric dams disrupt ecosystems both upstream and downstream; fossil fuels contribute to global warming, acid rain, and other forms of air pollution; and nuclear plants generate radioactive waste for which there is no disposal solution. The extraction, transport, and use of these fuels predictably lead to accidents—such as oil spills and radiation leaks—that can poison the environment for many years.

Broadcasting Unsustainability

People who are part of a healthy local economy have little need for an infrastructure that makes instantaneous global communication possible. But for the TNCs that dominate global trade, it is a necessity. Those communications networks enable corporations to coordinate production, distribution, and marketing among their numerous far-flung subsidiaries and to shift capital from country to country at the stroke of a computer key.

Like any large-scale infrastructure, communications networks have many direct ecological costs: launching communications satellites into orbit damages the ozone layer, for example, and transmitters create electromagnetic fields whose effects are still poorly understood. But there are far deeper environmental and social implications to this globally encompassing media infrastructure. Perhaps the most important of these is the way global media work to homogenize people's tastes, preferences, and desires worldwide.

One would expect food preferences to vary widely in different parts of the world, and that has indeed been the case throughout most of human history. People on the Tibetan plateau, for instance, enjoy yak-butter tea mixed with barley flour; in the New Guinea interior, taro and sweet potatoes are much-loved staples, with pork the occasional treat; among the Navajo of the American Southwest, maize and beans are preferred. Researchers are increasingly discovering that these diverse preferences make sense, both in terms of the products the local ecosystem can provide, and in terms of people's long-term genetic adaptations to those foods.

The global food system and the transnational corporations that dominate it, however, can thrive only if people have largely similar preferences everywhere and can be convinced to abandon their local foods for the monocultural products of the global economy. This is a clear and conscious goal of TNCs. Thus, a Campbell Soup Company annual report recently crowed over the prospect of replacing diverse homemade soups around the world with their own canned varieties:

> . . . most of the soup in Asia is still homemade, so our growth potential in this region brims with promise . . . In Mexico, our opportunities have been significantly broadened with the passage of NAFTA. With doors wide open to international trade, Mexico's 85 million people beckon as a highly attractive market . . .[24]

ISEC/John Page

How can people be induced to give up their diverse local foods for the uniform processed foods that TNCs produce and market? In the words of an advertising executive in China, they must be made to believe that "imported equals *good*, local equals *crap*".[25]

Perhaps the most insidious example of this is the successful campaign by Nestlé and other food corporations to convince Third World mothers that breast milk—the most ubiquitous of local foods—is inferior to the powdered version those companies sell.[26]

Creating consumer uniformity would be almost impossible without global media, which enable corporations to transmit direct advertising messages as well as glossy, unrealistic images of a Western, urban lifestyle. In the absence of direct experience, these glamorous images can be highly compelling to populations throughout the South, making the young in particular eager for the West's material trappings: not just bluejeans, sunglasses, and sports cars—but white bread, hamburgers, and bottled soft drinks as well.

Agribusinesses are well aware of the role the global media plays in homogenizing people's desires. As the CEO of the food conglomerate H.J. Heinz puts it, "once television is there, people of whatever shade, culture or origin want roughly the same things."[27] The corralling of more and more people into centralized urban areas is an important part of this process: it is much easier to reach urban populations with advertising images, and since they can no longer satisfy their own needs, urbanized people quickly find themselves dependent on a monetized economy dominated by TNCs.

These efforts at forging a consumer monoculture are effectively an attack on billions of people, mostly in the South, who must be made to reject their own individual, ethnic, and cultural identities. Corporations are more than willing to exploit the suffering this inflicts on target populations. The president of McDonald's Japan went so far as to suggest that

> the reason Japanese people are so short and have yellow skin is because they have eaten nothing but rice and fish for two thousand years. . . . [I]f we eat McDonald's hamburgers and potatoes for 1,000 years, we will become taller, our skin white and our hair blonde.[28]

Manipulating people around the world to "want roughly the same things" may be beneficial for food TNCs like McDonald's and Heinz, but for the global environment it is nothing short of disastrous. It should be clear by now that the earth can neither supply the resources for, nor absorb the wastes from, even a small percentage of the world's people living in suburban homes, driving sports utility vehicles, and eating frozen TV dinners.

Rather than perpetuating the hoax that the entire global population can and should someday live like affluent westerners—the implicit message of economic globalization—it would be far more socially and ecologically responsible to promote policies that reduce, rather than increase, unnecessary transport; that facilitate smaller-scale, decentralized markets, rather than huge centralized supermarkets; and that encourage greater dependence on locally available foods, rather than food transported from the other side of the planet.

Global Food and Climate Change

Is the global food system really a major contributor to greenhouse gas emissions? Yes, although the exact magnitude of its contribution is all but impossible to quantify. In the United States, for example, food transport totaled some 566 billion ton-miles in 1997, accounting for more than 20 percent of all commodity transport.[j] Based on that figure and emission rates for various modes of transport, a conservative estimate for CO_2 emissions directly attributable to in-country food transport is some 120 million tons every year.[k]

But this is just the tip of the iceberg. For one thing, that figure does not account for the transport of foods beyond the borders of the United States. In 1998, 172 million tons of food were shipped to and from the United States.[l] It is likely that this cross-border transport contributes as much or more CO_2 as food that never leaves the country: for example, an avocado grown and harvested in Chile, air-shipped to California, and then trucked to a supermarket in Arizona accumulates most of its transport miles before it ever reaches America's borders. And since travel by air is responsible for more CO_2 per ton-mile than any other mode of transport, the miles traveled outside the country would be the most ecologically damaging.

What's more, these calculations account only for the transport of food itself, and not for all the commodities that support the global food system. In the United States, the transport of pesticides and fertilizers—mainstays of industrial agriculture—accounted for an additional 46 billion ton-miles in 1997.[m] Other commodities required by the global food system include plastic and paper for packaging; refrigerants and other chemicals for food preservation; oil and gasoline to fuel industrial farm equipment; and the trucks, planes, barges, and rail cars needed for food transport.

Another toll on the atmosphere comes from the construction of supermarkets and distribution centers as well as the miles of roads and acres of parking lots on which America's globalized food system depends: cement manufacture, for example, is a significant contributor to atmospheric CO_2 while road construction often means the loss of woodlands that previously served as valuable CO_2 sinks. Global foods also demand large amounts of electricity for refrigeration and food processing, a demand often filled by fossil-fuel burning power plants, which are among America's worst greenhouse gas emitters.

Such a list could go on and on, but it would be incomplete if it did not include the CO_2 cost of bringing American consumers to the supermarkets where food from the global system is sold. The spread of these large centralized supermarkets is requiring people to travel further than ever before—usually by car—to purchase food.

Accounting for all these direct and indirect costs, America's dependence on the global food system is clearly responsible for a significant portion of the nation's greenhouse emissions. But while the United States is the world's biggest contributor to global warming, it is not alone. In 1998, for example, UK food and agricultural products accumulated over 33 billion ton-miles of transport within the country's borders, directly adding some 13 million tons of CO_2 to the atmosphere.[n] Meanwhile, more than 44 million tons of food were imported or exported, adding still more greenhouse gases.[o] Even worse, air shipments of food to and from the United Kingdom are on the rise, with imports by air doubling between 1980 and 1990.[p] Since CO_2 emissions per ton-mile are nearly 6 times higher for air transport than for road transport and almost thirty times higher than for rail,[q] it is likely that global foods are responsible for an increasing proportion of the United Kingdom's total greenhouse gas emissions.

A similar story is unfolding around the world: as countries everywhere increase their dependence on the global economy, CO_2 emissions are steadily rising and the climate is becoming ever more unstable. Human-induced global warming is arguably the most pressing ecological issue of our time, and it is little short of madness to exacerbate the problem by promoting the unnecessary transport of food—one of the few products that people need every day.

3

The Ecology of Food Production

An enduring agriculture must never cease to consider and respect and preserve wildness. The farm can exist only within the wilderness of mystery and natural force. And if the farm is to last and remain in health, the wilderness must survive within the farm.

—Wendell Berry, *The Unsettling of America*

THE HIGH ECOLOGICAL COSTS of global food are not only products of its distribution and marketing: *producing* for the global market is also highly damaging to the environment. And once again, local food is relatively benign by comparison.

One reason for this stems from the numerous levels of diversity inherent in local food systems. Local foods tend to differ from place to place, in direct relation to differences in climate, geography, and natural resources. Similarly, local food production involves a wide range of cultivation methods, as each locale's unique ecological and cultural conditions are allowed to determine appropriate farming practices.

Wherever people's needs are largely supplied by a local food system, the farms in that region are themselves more diverse. A region with nothing but monocultures—like the several-thousand acre wheat fields of Kansas—can produce huge amounts of a single crop for global markets, but people need more than just one or two foods. Farmers who supply local markets therefore have strong incentives to diversify their production.

Local food systems also support more diversity within individual crop species. For millennia, seed-saving farmers have selected plants for certain traits, including their success in local microclimates and soil types. Agricultural biodiversity has steadily multiplied as a result. The more than 17,000 distinct varieties of wheat that exist today are a product of many centuries of careful seed selection in varied ecosystems.[1]

When farms are small in scale—and particularly when they are farmed organically—they also enable a wide range of non-food species to coexist

Michael Ableman

within the farm system. Hedgerows, woodlots, pastures, and fallow land become nurturing habitats for numerous wild plant and animal species, thereby helping to maintain a region's overall biodiversity. In some cases, the farm itself mimics the wilderness, as in the traditional forest gardens of the Tamil Nadu highlands in southern India. These gardens produced a fantastic array of fruits, nuts, berries, roots, and edible leaves while relying on the forest's indigenous species—including microorganisms, insects, wild animals and "non-productive" plants—to maintain the garden's balance and health.

Destroying Diversity

Production for the global market, on the other hand, effectively precludes diversity. What a farm produces is not determined by local conditions but by the requirements of a global marketing system that prizes standardized products, extended shelf life, and the capacity to withstand long-distance transport. This has led to an agriculture that is highly specialized, growing a shrinking number of varieties. Wild nature, meanwhile, is systematically excluded from industrial farmland, adding to the pressure on non-agricultural plant and animal species whose habitats have already been whittled down by encroaching development. Studies in Germany, for

ISEC/Steven Gorelick

instance, have shown that industrial farming is that nation's leading contributor to biodiversity loss, with over 500 plant species alone endangered or extinct as a result of agricultural practices.[2]

As farmers specialize production in fewer and fewer varieties of the one or two crops they grow, the planet's agricultural biodiversity steadily erodes. In the United States today, almost three-quarters of potato production comes from just four closely related varieties; 76 percent of the nation's harvest of snap beans comes from just three strains; and 96 percent of pea production comes from just two pea varieties.[3] The corn industry is so dependent on inbred lines of hybrids that one seed company official admitted that "the corn seed industry is probably working from the narrowest base in history."[4] If genetically engineered crops are allowed to dominate the fields of industrial farmers, agricultural diversity will deteriorate still further.

On-farm vs. Off-farm Inputs

The lack of diversity on industrial megafarms has troubling implications. As a farm's diversity declines, it becomes much less stable and resilient. Insect pests and blights that favor particular plants can quickly spread from one to another, making monocultures much more susceptible to devastation. While farmers practicing diverse agriculture are apt to lose only a

USDA/Dave Hein

small percentage of any crop to pests, diseases, or weed invasions,[5] a farmer with a monoculture can lose virtually everything. Fearing financial hardship or even ruin, large-scale farmers understandably try to minimize the damage by using pesticides, herbicides, fungicides, and other chemicals—especially since these are the methods most heavily promoted by agribusinesses, development experts, and government farm agencies.

But these solutions are shortsighted at best. Most pesticides kill a broad range of insects, and their use only worsens the problem in the long run by eliminating the target pest's natural predators. Meanwhile, pests inevitably develop resistance to the chemicals that target them, which means that either heavier doses or new pesticides must eventually be deployed. The use of these chemicals thus creates a vicious cycle: the more they are used, the more serious the problem of pests becomes, and the greater the incentive to use more, and stronger, chemicals.

These expensive inputs—part of the technological treadmill that diverts farmers' incomes into the hands of corporate agribusinesses—are highly damaging to nearby ecosystems. Since pesticides and herbicides kill far more than the few insect or weed species they target, their use all but guarantees that the surrounding ecosystem will be less complex, less diverse, and less stable.

Pesticides are by no means the only chemical input that industrial farmers use. While small-scale, diversified farmers can feed the soil by using composted manure and other organic matter from the farm, farmers planting large-scale monocultures tend to use chemical fertilizers to feed the plants directly. The use of these inorganic fertilizers has increased dramatically since the end of World War II. In England and Wales, for example, farmers used six times more inorganic nutrients on winter wheat in 1995 than they used in 1945.[6] In the US, the use of nitrogen fertilizers rose from about 300 million tons per year in 1945 to nearly 10 *billion* tons by 1985.[7] Similar increases are the norm worldwide.

The chemicals applied to fields are never used up entirely by the target crops, and much is lost to the environment as agricultural pollution. Some 30 to 80 percent of nitrogen in fertilizers and small but significant quantities of pesticides are lost directly to the environment, where they contaminate water, food, fodder, and the atmosphere.[8] These chemicals can travel widely from the fields in which they are sprayed: even polar bears and other arctic animals, as well as marine mammals such as whales and dolphins, contain detectable amounts of pesticides in their fat.[9]

The pollution problems arising from large-scale, intensive agriculture have been widely documented. Pesticides can poison soil, rivers, and streams, and often harm wildlife. Many species—including the bald eagle and California condor—are endangered today because of pesticide spraying many years ago. Pesticides are also implicated in the decline of frog populations worldwide and in the increasing numbers of deformed frogs seen in the United States.[10]

Nitrates and phosphates from chemical fertilizers, meanwhile, contribute to algal blooms in surface waters, causing eutrophication and fish deaths.[11] A large-scale example of this phenomenon is the huge area in the Gulf of Mexico where there is little marine animal life because of low oxygen levels. This 20,000 square kilometer dead zone is a consequence of agricultural run-off from US farms, that enters the gulf from the Mississippi River.[12]

Another ecological cost of these inputs occurs during their manufacture. One of the worst industrial accidents in history occurred in Bhopal, India in 1984, when gases leaking from a Union Carbide pesticide plant killed more than 6,000 people within the first week and more than 10,000 in the years since, as well as doing untold damage to the local environment.[13] There have been thousands of smaller-scale spills and accidents at chemical input manufacturing facilities around the world, each taking its toll on the environment.

The most important input used in agriculture is, of course, water. Agricultural areas where rainfall is inadequate, unreliable, or absent altogether require some form of irrigation. When conducted on the small scale typical of local food systems, irrigation has little environmental impact beyond the farmer's fields. In the Himalayan region of Ladakh, for example, hand-constructed irrigation channels divert glacial meltwater into terraced fields that have remained highly productive for many centuries.[14]

When used on the scale required by the global food system, however, irrigation infrastructures and the large dams they require have a major impact on the environment. As water levels rise and water temperature and sediment content are abruptly altered, the riparian ecosystem above and below the dam is severely disrupted. Large dams also flood large expanses of land that had previously served as human, animal, and plant habitats. In India, for example, the huge Tehri Dam is set to flood 68,000 acres of prime farmland and displace 100,000 people from their homes.[15]

Dams can also destroy traditional agricultural systems that rely upon natural flood cycles for their viability. The construction of the Aswan High Dam in the south of Egypt, for example, ended the annual floods on which agriculture in the Nile Valley had depended for millennia. The nutrient-rich sediments that were deposited on agricultural land every year as the floods subsided are now trapped behind the dam. Egyptian farmers must now use chemical fertilizers, while the dam itself is rapidly becoming silted up.

Throughout history, excessive reliance on large-scale irrigation networks has resulted in salinization of the soil, and the eventual abandonment of once-fertile land. The Centre for Resource and Environment Studies in Australia estimates that 20 percent of all irrigated land—more than 112 million acres—are already salt-affected. Each year, 5 to 8 million acres become so badly salinated that they must be abandoned.[16]

Integrated Livestock vs. Factory Farms

People who have visited a diversified, small-scale farm often observe that everything seems to have multiple uses, and that almost nothing goes to waste. This is particularly the case when animals are part of the farm mix. Crop residues can feed horses, pigs, and goats; the chaff remaining from grain crops can be used for animal bedding or fodder. Table scraps and vegetable waste can be fed to chickens and pigs. Even bits of leaves, branches, and bark that remain from gathering firewood can be eaten by goats or horses. The animals, meanwhile, can produce meat, milk, eggs, hides, and fiber for clothing, and some can power farm equipment. When allowed to range freely, chickens, geese, and other fowl eat large numbers of harmful insects and slugs. Many ruminants can graze on untillable land, thus expanding the farm's productive area. And on these farms, the manure produced is highly valued, since it can be used to maintain the fertility of fields and pastures.

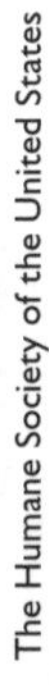
The Humane Society of the United States

Monocultural production for distant markets, on the other hand, separates animal husbandry into a large-scale, intensive activity completely delinked from other farm processes. Thousands or even millions of animals are caged or penned within these factory farms, and tons upon tons of manure pile up and must be disposed of. A cattle feedlot with 20,000 cows—modest by industry standards—produces as much sewage as a town of 320,000 people.[17] While small farms find an immediate use for their manure, the massive amounts of manure produced on factory farms become a serious source of pollution, even affecting regions far beyond the immediate farm vicinity. Not only are streams, rivers, and groundwater at risk, but the air can be polluted as well: it is estimated that 30 percent of acid rain in the Netherlands, for example, is a product of the country's industrial livestock operations.[18]

Meanwhile, feeding all these penned animals requires tremendous quantities of grain, usually grown in vast monocultures that need, among other inputs, chemical fertilizers. Thus, the elegant simplicity of the small farm, which can produce much of the feed its animals need on fields fertilized by the animals' own manure, is replaced by a globalized system in which manure is a liability in one place and soil fertility is in short supply elsewhere. As Wendell Berry has succinctly put it, "The genius of . . . farm experts is very well demonstrated here: they can take a solution and divide it neatly into two problems."[19]

Life in the Soil

On more diversified farms, it is possible to use a wide range of strategies to maintain soil fertility. In some cases, as with traditional shifting agriculture, land is allowed to return to the wild for long periods after a few years of cultivation; in other cases, farmers replenish soil nutrients by incorporating food waste, crop residues, animal manure, and even human night soil. In organic agriculture, the soil is viewed as a living resource that must be properly nourished in order to produce healthy crops for the long term. And indeed, healthy soil is teeming with life forms, all of which play a vital role in maintaining good soil structure and in breaking down organic matter.

When you produce tons of a single crop using massive machinery, it is difficult to view the soil as anything but an anchoring medium for plant roots and a receptacle for water and chemical fertilizers. Those chemicals, as well as pesticides, herbicides, and fungicides, effectively end the symbiotic relationship between soil and plant by killing soil organisms, thus transforming the soil into a dead, toxic substance.

Many other industrial practices lead to degraded soil. The numerous trees, shrubs, and hedgerows that help keep soil in place on small-scale farms are uprooted on industrial farms to make way for large machinery. Crop residues are often removed from fields after harvesting, thereby exposing the soil to leaching from rain and desiccation from the sun. Large machinery tends to compact the soil, reducing its ability to absorb water. And the chemicals and cultivation methods characteristic of large-scale monocultures alter soil consistency, making it more prone to water and wind erosion.

As a result, topsoil is being lost at an alarming rate. It is estimated that five pounds of topsoil are lost for every pound of grain harvested in Iowa; in eastern Washington, each pound of grain costs *twenty* pounds of topsoil.[20] Eventually, productive agricultural land is destroyed. The eroded soil, meanwhile, severely disrupts watercourses, causing flooding that damages natural resources and human settlements.

Is Organic Enough?

Today, what is meant by organic food has become a subject of heated dispute in many countries. Giant business interests are influencing its guidelines and definitions, to allow them to take better advantage of the burgeoning market for organic foods—a market that agribusinesses and food corporations have already invaded. Huge food corporations are buying up small independent producers and natural food businesses. Many organic farms are in reality large monocultures, albeit ones that refrain from the use of chemical inputs. And while most organic produce was until fairly recently sold directly to consumers or through health food shops and other small retailers, today most organic food is sold through supermarket chains. In the United Kingdom, for example, 69 percent of all organic produce is sold by supermarkets.[a]

The possibility that agribusinesses and food corporations will hijack the organic movement is very real. In 1999, for example, the US Department of Agriculture (USDA) attempted to impose a weak, corporate-friendly set of organic standards on the nation. Among other flaws, the proposed rules would have allowed foods labeled as organic to be grown from genetically modified seed, fertilized with chemically tainted municipal waste, and sterilized by irradiation—techniques considered acceptable within the global food system, but consistently repudiated by organic farmers. Although the USDA backed away from these controversial rules in the face of outrage from consumers and farmers alike, the integrity of the organic label in the United States is still under threat.

In the ongoing battle over the meaning of organic, agribusinesses prefer to define the term as narrowly as possible. Their focus is on a short list of prohibited farming practices, particularly the use of chemical inputs such as pesticides, fungicides, herbicides, and artificial fertilizers, since those requirements can be met without challenging the premises that underlie global food. Issues of nutritional value, waste, animal welfare, and environmental impact are largely outside this narrow focus.

Growing without chemical inputs does minimize the problems of pesticide pollution and residues on food, but when so-called organic foods are produced in large-scale monocultures and transported thousands of miles, many of the other costs of the global food system remain: the rapid loss of biodiversity, the destruction of independent shops by corporate-owned hypermarkets, the undermining of local economies and communities, and the environmental costs of long-distance transport. The food, meanwhile, is more likely to be packaged, processed, and preserved to increase its shelf life, thus reducing the health benefits of eating organic food.

Furthermore, large-scale organic production actually increases the already overwhelming pressures on more sustainable smaller-scale producers. The local food initiatives that enable small producers to thrive can easily be flooded with foods from large-scale organic producers, who can take advantage of the hidden subsidies of the global economy to sell their products at artificially lowered prices.

Conversely, production oriented toward local markets is far more conducive to organic methods. Since local food needs and preferences are themselves diverse, local marketing encourages agricultural diversity, thereby reducing the threat of crop devastation by blights and pests and making it much easier for farmers to avoid the use of chemicals.

Moreover, locally oriented organic production has economic benefits for both the farmer and the consumer. Since fewer inputs are needed, and since fewer processors, distributors, and other middlemen are taking a cut, closer links between organic farmers and consumers simultaneously lower the price of healthy food for the consumer and raise the amount that goes to the farmer.

ISEC/Steven Gorelick

Farming Organically

One of the guiding principles of most organic farmers is that healthy crops demand healthy soil, and so they emphasize maintaining soil fertility and good soil structure. This is achieved in a variety of ways. Cover crops—particularly nitrogen-fixing legumes—may be grown during the off-season or as part of a rotation, and then tilled into the ground to increase organic matter in the soil. Other organic materials such as kitchen scraps, crop residues, and animal manure are composted and added to the soil to build humus and add nutrients.

Crops are usually rotated (i.e., planted in different spots from one growing season to another) to keep any one crop from overburdening the soil with its particular nutritional demands. Intercropping—growing multiple crops among each other in the same space—ensures the efficient use of the agricultural plot, helps suppress weeds, and allows sympathetic species to benefit each other by their proximity. Pests find it more difficult to spread through mixed crops, especially if some of the plants repel the pests of other plants, or play host to their predators. Biodiversity is emphasized, both in the number of species cultivated and in the complex ecological system that is created on the farm. This diversity enhances the stability of the whole system. A recent study of paired organic and nonorganic farms across England revealed that organic farms have lower nutrient losses to air and water, and significantly more butterflies, predatory spiders, and floral diversity.[b]

While the global market pushes farmers to specialize in producing a single crop for export—effectively demanding monocultures that make it difficult or impossible to incorporate organic practices—diversified organic farms contain so much biodiversity that pests are not as serious a problem, and doing without pesticides is not difficult.

In addition, the drastically reduced need for external inputs means much less fossil fuel use. Organic systems have been shown to require 60 percent less fossil fuel per unit of food produced.[c] However, it is primarily on the smaller, less mechanized, and more biodiverse farms that major energy savings are possible: large-scale organic monocultures generally involve the use of almost as much energy as conventional farms. And of course, locally oriented organic food also requires much less fossil fuel to get the produce to market.

An Organic Resurgence

Organic farming has expanded in the industrialized West in recent years, as food scares and increasing concerns about the genetic manipulation of food lead consumers to seek out more trustworthy food sources. In the United States, demand for organically grown foods has grown by more than 20 percent per year for the last several years (see figure 3.1). In Europe, the amount of agricultural land devoted to organic cultivation has increased rapidly in the past decade (see figure 3.2). In Austria, where government policies support organic farming, more than 10 percent of agricultural land is now cultivated organically.[d]

Although the growing popularity of organic foods is a positive development, it is vitally important that organic does not become just another brand name or the intellectual property of the global food industry. Environmentally conscious and health-minded consumers therefore need to look not only for the organic label, but

Figure 3.1: Rise of Organic Food Industry in US

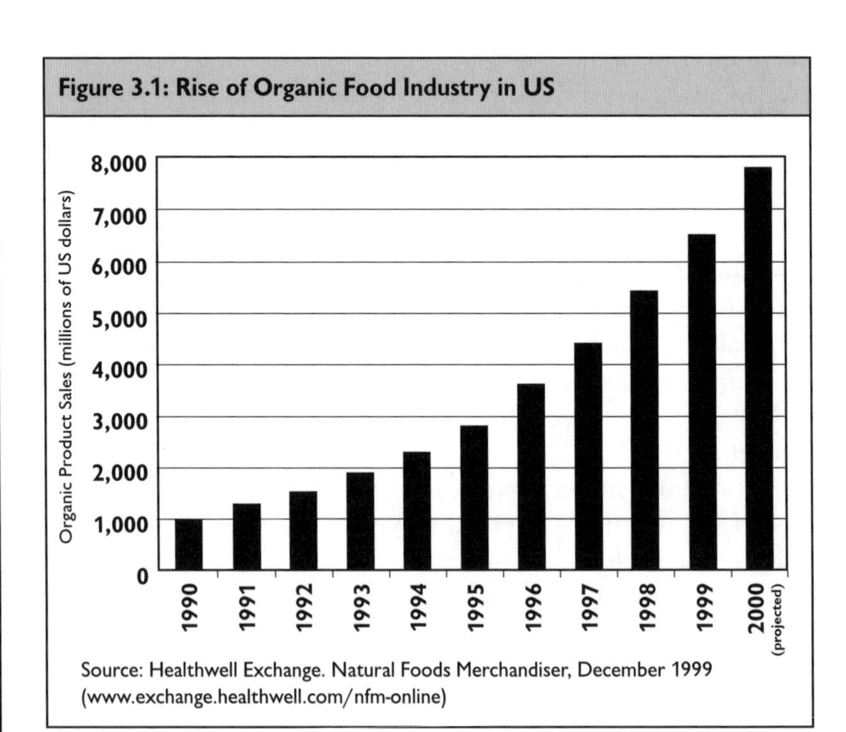

Source: Healthwell Exchange. Natural Foods Merchandiser, December 1999 (www.exchange.healthwell.com/nfm-online)

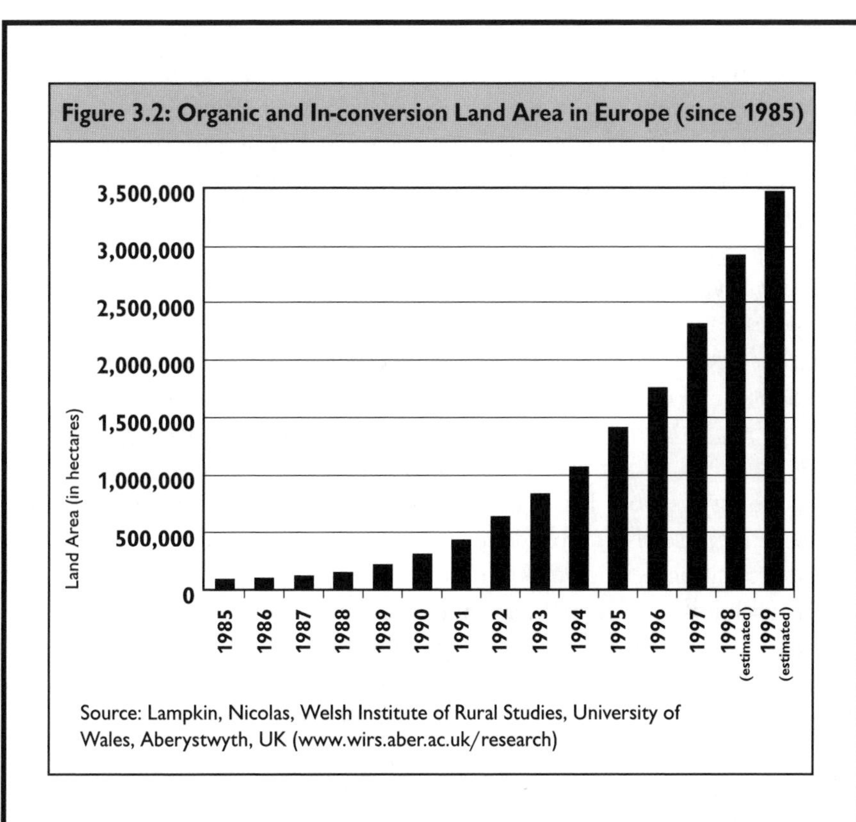

Figure 3.2: Organic and In-conversion Land Area in Europe (since 1985)

Source: Lampkin, Nicolas, Welsh Institute of Rural Studies, University of Wales, Aberystwyth, UK (www.wirs.aber.ac.uk/research)

should seek out locally grown organic foods whenever possible. At the same time, communities and activist groups need to pressure their governments to adopt guidelines that protect "organic" against co-optation by large agribusinesses and food corporations, while supporting appropriate subsidies and regulatory structures that make small-scale, locally oriented organic agriculture more possible.

Local Adaptation vs. Genetic Engineering

Since the beginnings of agriculture, farmers have selected for traits that make the most sense within their own particular environment, thereby providing almost every local food system with a remarkably broad range of locally adapted plant varieties and animal breeds. Indigenous farmers in the Andes, for example, cultivate some 3,000 different varieties of potatoes.[21] On the island of Java, small farmers cultivate more than 600 different crop species in their gardens.[22] Much of the food we eat today ultimately depends on the careful work, over many centuries, of farmers like these.

Now, however, an entirely new method for creating agricultural varieties has been developed. Rather than selecting for particular traits among plants and animals that have proven themselves in nature over centuries, biotechnology enables scientists to select traits in the laboratory. In many cases, scientists carry genetic material across entire species or phyla boundaries, bypassing reproductive constraints and creating varieties that could never have evolved in nature, even with the guiding hand of a skilled breeder. Fish genes have been implanted into tomatoes, and human genes into fish. There has even been research into engineering such labor-saving traits as a featherless chicken that won't have to be plucked.[23]

Leaving aside the ethical implications of manipulating the genetic basis of life, this technology may have severe ecological repercussions. For one, biotechnology is now being used to *increase* the use of many pesticides: Monsanto sells seeds that produce crops engineered to tolerate heavier doses of its best-selling herbicide, Roundup[24]; Aventis markets similar seeds, but for use with its own Liberty herbicide; and Cyanamid has produced seeds to be used with its Pursuit and Odyssey herbicides.

Some biotech varieties are engineered to kill the insect pests that eat them. But studies have shown that, like chemical pesticides, these plants can also harm nontarget insects. Cornell University researchers sprinkled milkweed—a favored food of monarch butterfly larvae—with pollen from Bt corn, which is genetically engineered to produce its own pesticide. Though Monarch butterflies are not targeted by the genetically engineered pesticide, their larvae were nonetheless killed when they fed on the milkweed.[25]

Perhaps most disturbing of all is the problem of genetic pollution, whereby crops or wild plants are accidently fertilized by a nearby biotech crop. Although proponents of genetic engineering have claimed that such cross-fertilization would be rare, it has not turned out that way. For example, Starlink, a transgenic corn variety planted on less than 1 percent of America's corn acreage, managed to contaminate the seed corn of more than eighty seed companies. Meanwhile, the appearance of herbicide-resistant "superweeds"—the result of genetic pollution from transgenic crops—has already been documented.[26]

It is now clear that genetic pollution can travel significant distances. Research in the remote mountainous region of Sierra Norte de Oaxaca has shown that even some of Mexico's native varieties of corn have been contaminated by transgenic DNA, even though no genetically engineered corn was intentionally planted within sixty miles. "I repeated the tests at least three times to make sure I wasn't getting false-positives," said David Quist, lead author of the study. "It was initially hard to believe that corn in such a

© Nick Cobbing/David Hoffman Photo Library

remote region would have tested positive."[27]

The fact is, so much genetic pollution is occurring that there is a danger that farmers—even organic growers—will soon be unable to find seed that is *not* tainted with engineered genetic material. "We have found traces in corn that has been grown organically for 10 to 15 years," the head of an organic bread and cereal company in British Columbia said. "There's no wall high enough to keep that stuff contained."[28] According to Farm Verified Organic, a US-based organic certifier, genetic pollution of corn, canola, and possibly soybeans is already so pervasive, "it is not possible for farmers in North America to source seed free from it."[29]

Many people now believe that this was the strategy of the agricultural biotech companies all along: rather than trying to contain genetic pollution, the industry was counting on contamination so pervasive that any debate over engineered crops would be rendered irrelevant. According to biotech critic Jeremy Rifkin, "They're hoping there's enough contamination so that it's a fait accompli."[30]

But far from convincing us that we might as well embrace engineered foods, the frightening rate of genetic pollution tells us that we should call a halt to the spread of this technology *immediately*. There are still many crops for which transgenic seeds have yet to be developed or planted, and many countries are still free of genetic pollution. Should we contaminate these gene pools as well?

Now, genetically-engineered insects are coming out of the biotechnology pipeline as well, with the first field release already underway.[31] Releasing transgenic insects into the wild—an irreversible action—poses ecological risks no scientist can hope to understand, much less dismiss. If the corporate scientists were wrong about the likelihood of genetic pollution, what else are they wrong about?

Biotechnology will only add to the ecological costs of the global food system. The widespread adoption of genetically engineered crops would mean that the locally adapted varieties used by farmers for centuries, each tailored to a region's unique ecological conditions, would give way to a few select varieties designed and owned by large agribusinesses. This would only accelerate the already alarming rate at which agricultural biodiversity is being lost; further weaken farmers' control over food production; and further homogenize diets, cultures, and rural landscapes—all at potentially great ecological cost.

Despite the industry's public relations spin—which attempts to portray genetic engineering as a boon for farmers, the environment, and the poor—biotech is at best a temporary and desperate techno-fix, one that does nothing to reverse the destructiveness of the global food system. Instead, its main focus will be on maintaining and expanding that system—by enhancing a monoculture's ability to resist pests, by limiting damage to vegetables during long-distance transport, by creating new colors to improve the supermarket appeal of vegetables, and so on.

Trusting in high technology and profit-driven corporations to restore and maintain the health of the planet would be an act of profoundly misplaced faith. But reducing the scale of our economic systems, starting with those that produce and deliver our food, would allow diversity to flourish once again and could set us on the path to environmental health.

4

Food and Health

The only time I felt truly comfortable about the food I put on my table was when I lived on the farm and grew most of my own . . . Now, I live in an apartment in the city, and am dependent on nameless, faceless strangers to grow, process and ship my food. It seems as if unethical and unsafe practices grow in direct proportion to how far we have lost the trail of accountability. So I don't always trust them to put my family's best interest over concern for their bottom line. I don't like feeling helpless, as if every trip to the grocery is a crap shoot.

—Vicki Williams, columnist, *USA Today*

IF ONE CONSIDERS some typical modern foods—hamburgers laden with growth hormones, vegetables laced with pesticides, soft drinks full of refined sugar, and foods too numerous to mention whose color and taste have been artificially enhanced by manufactured chemicals—one could easily imagine that the goal of the global food system is simply to provide the global *health care* system with more customers. Local food systems, on the other hand, are not only healthier for the environment, they provide people with healthier food as well.

Fresh Is Best

Local systems excel at providing fresh food, and health practitioners of every stripe agree that fresh food is the most nutritious. Some nutritionists have even determined that the best nutrition of all comes from foods that are in season in one's locale. Since the vitamins in almost any food are gradually lost from the time of harvest, even fresh foods from the global system are usually less nutritious than local foods, because they may have been harvested days or even weeks before reaching the kitchen table. Tomatoes, for example, are often picked green and hard so that they can survive mechanical harvesting and long-distance transport, and then ripened in rooms pumped full of ethylene gas, which artificially initiates the ripening process. Tomatoes like these are much less flavorful and nutritious than the ripe tomato from a local farm, plucked from the vine and eaten the same day.

Foods for the global market are bred for monocultural growing conditions and the ability to be transported long distances, rather than for nutritional content. Another high priority is visual perfection. Decades of agribusiness and supermarket advertising, combined with numerous senseless regulations, have persuaded people that fruits and vegetables must conform to narrow standards of size, shape, and color. Customers expect to find only bright red, unmarred apples, potatoes that are properly shaped and without blemish, and carrots that are large, straight, and orange. Most western consumers are now so disconnected from agricultural reality that heirloom varieties of unusual shape or color are not considered to be real foods at all. And food grown in living soil where insects are allowed to survive—sometimes leaving their mark on the produce—is considered substandard, even though it is likely to be better tasting and more nutritious than its more perfect-looking industrial cousin.

Biotech varieties are no exception to the rule among global foods. Thus, despite the inflated claims about the virtues of genetic engineering, the varieties that have reached the supermarket so far have not been improved nutritionally. Roundup Ready products, for example, are engineered to survive herbicide drenchings; Flavr-Savr tomatoes are designed to sit on supermarket shelves for long stretches of time without rotting; and Bt corn and Bt potatoes have been engineered to contain a potent pesticide—not extra nutrients—in every cell. Although so-called Golden Rice has been engineered to contain extra amounts of vitamin A—and is being touted as a cure for a form of blindness caused by vitamin A deficiency—its main beneficiary thus far has been the biotech industry, for which it has served as a much-needed public relations vehicle. (See box, page 53).

Chemical Stews

Global foods tend to undergo a great deal of processing, which destroys nutrients. Some highly refined products such as white flour, white sugar, and white rice have had most of their nutritional content stripped away. Since processing can also remove much of the taste and color from food, the global food industry often compensates by adding artificial flavorings and colorings. In some cases, these chemicals are used simply because they are cheaper than real flavorings and spices—as when real vanilla is replaced by vanillin, a chemical substance that approximates the flavor that comes from vanilla beans. Chemical preservatives are also deployed, to add to the extended

The Golden Rice Hoax, by Vandana Shiva

To many cultures in Asia, rice is life itself. And this is why the ongoing corporatization of rice varieties is such a tragedy.

Rice evolved as a food source in Asia, in many and varied forms. Recently though, the globalization and corporatization of agriculture has had serious effects on that diversity. India, for example, had nearly 200,000 rice varieties until that rich genetic diversity was destroyed by the chemicals and machines of the Green Revolution. The Green Revolution's scientists "built" India new rice varieties to replace the thousands it destroyed; but in doing so, they also created forty new insect pests and twelve new diseases for rice farmers to cope with. The net result: a worse life for farmers, and fewer varieties of rice.

As the Green Revolution miracle fades, the world's technocrats are preparing its second wave: genetic modification. One of the first products we are seeing is Golden Rice, which is being proclaimed as a miracle cure for blindness. More than $100 million has been spent over ten years to produce this transgenic rice at the Institute of Plant Sciences in Zurich. The Zurich team introduced genes taken from daffodils and bacteria into a rice strain, to produce a yellow rice with high levels of beta-carotene, which is converted to vitamin A within the body.

Now, plans have been drawn up for a transfer of Golden Rice to India. And for what? Vitamin A rice is likely to fail in preventing blindness, since it will meet less than 1 percent of the required daily intake. Ninety-nine percent of vitamin A will still have to be provided from alternatives which already exist, such as green leafy vegetables and fruits—coriander leaves, curry leaves, drumstick leaves, amaranth leaves—staples of the Indian diet.

In fact, as ever with such miracle technologies, Golden Rice is based on a false premise. The destruction of biodiversity by industrial agriculture is a primary cause of today's vitamin A deficiency across rural India, and it is only through rejuvenating biodiversity on our farms that we can solve problems of vitamin deficiency and malnutrition.

In spite of all the hype about Golden Rice, it will aggravate far more problems than it alleviates. Genetically engineered vitamin A rice is part of industrial agriculture, and as such it requires intensive inputs. It will also lead to major water scarcity, since it is a water intensive crop and will displace more water-prudent sources of vitamin A. And while the gene giant agribusinesses are reaping public relations benefits by granting royalty-free licenses for Golden Rice to Third World countries, the patents and ownership of the rice remain firmly in corporate hands.

No corporation can reproduce the amazing diversity of rice that nature and peasants have evolved in partnership over millennia—rice that grows up to eighteen feet to survive floods, rice that is salt- and drought-tolerant, rice that is aromatic and therapeutic. This diversity, and the knowledge and culture it embodies, is the real basis for health and for future food security. We must fight to keep rice free—in all its amazing diversity. Because on the freedom of rice depends the freedom of millions of Third World farmers.[a]

Vandana Shiva is director of the Research Foundation for Science, Technology and Ecology, in New Delhi, India, and a prominent environmental activist.

shelf life global foods require.

Local foods often contain no chemical additives, since they are less likely to need processing. And because of the prevalence of small, diversified, organic farms in local food systems, these foods are less apt to contain residues of pesticides, herbicides, and other toxic agrochemicals.

Although these chemicals now routinely turn up in our food and water, they are very recent in human evolutionary history, and our defenses are therefore unprepared to protect us from them. They can cause cancer, birth defects, immune system breakdown and neurological damage, and can interfere with normal childhood development.[1] Some of these chemicals are endocrine disrupters and have been implicated in the early onset of puberty so prevalent in the industrialized world. Studies have even indicated a correlation between aggression and exposure to pesticides.[2] The chemical fertilizers used in industrial agriculture also pose a health problem: nitrates in water, for example, have been linked to blue-baby syndrome in infants, birth defects, and cancer of the gastrointestinal tract.[3]

The health of farmworkers is seriously compromised by their exposure to agricultural chemicals on the job. According to a United Nations study, from 20,000 to 40,000 farmworkers die each year from pesticide exposure.[4]

USDA/Tim McCabe

Another study indicates that as many as 300,000 farmworkers in the United States alone suffer from pesticide-related illnesses.[5] But one doesn't need to be a farmworker or even live near a farm to be exposed to these toxic compounds. As Peter Montague of the Environmental Research Foundation points out, "Tens of millions of Americans in hundreds of cities and towns have been drinking tap water that is contaminated with low levels of insecticides, weed killers and artificial fertilizers. They not only drink it, they bathe and shower in it, thus inhaling small quantities of farm chemicals and absorbing them through the skin."[6]

If anything, Montague may have understated the extent of the problem. A recent survey by the US Environmental Protection Agency found that 80 percent of adults and 90 percent of children in the United States have measurable concentrations of the pesticide chlorpyrifos in their urine.[7]

Although agribusinesses insist that all of these chemicals have been tested for safety, they are not tested in the multiple combinations to which people are routinely exposed, nor are they tested over the long periods of time that would be necessary to fully understand their effects. Determining the so-called safety of individual chemicals is an all but meaningless exercise, since people in the industrial world are effectively immersed in a stew of such chemicals—arising not only from industrial agriculture but from fossil fuel use and manufacturing processes as well. In the United States, for example, roughly 1,000 new chemicals are marketed each year, adding to the 70,000

USDA/Doug Wilson

already on the market. A study in the journal *Science* points out that testing the commonest 1,000 toxic chemicals in unique combinations of three would require approximately 166 million experiments. Even if just one hour were devoted to each experiment and one hundred laboratories worked twenty-four hours a day, seven days a week, the process would take more than 180 years to complete.[8] Needless to say, no one is planning to conduct those tests.

In any case, the proven health hazards of a particular agricultural chemical are no guarantee that its use will be prohibited. The herbicide Atrazine is a known carcinogen whose use has been banned in seven European countries. Nonetheless, it is perfectly lawful to use it on fields throughout the United States.[9] Unfortunately, Atrazine is not the exception, but the rule: the US government agencies that regulate agricultural chemicals allow at least thirty other pesticides classified as either "definitely" or "probably" carcinogenic to be used on US crops.[10] Even chemicals commonly advertised as totally benign to humans can turn out to be harmful. For example, a 1999 study in the journal of the American Cancer Society linked exposure to glyphosate—the active ingredient in the herbicide Roundup—to non-Hodgkin's lymphoma, a form of cancer.[11] In 1998, more than 112,000 tons of glyphosate were used worldwide.

In addition to the inputs chemical farmers intentionally pour on their crops, there are numerous pathways by which global foods can be unex-

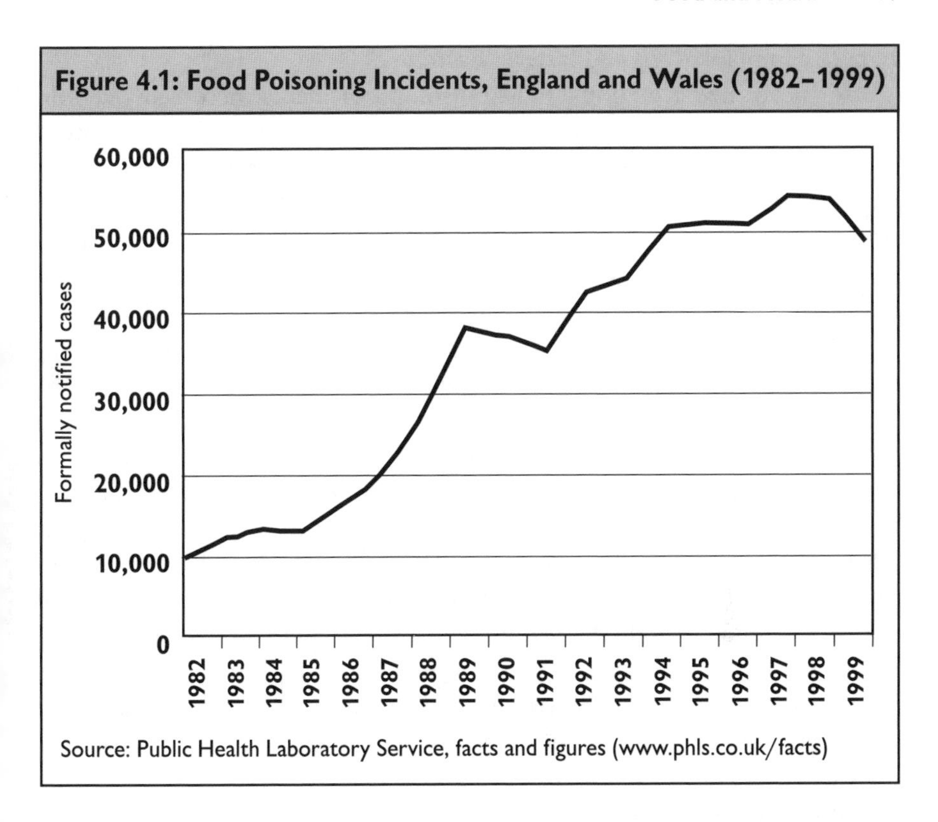

Figure 4.1: Food Poisoning Incidents, England and Wales (1982–1999)

Source: Public Health Laboratory Service, facts and figures (www.phls.co.uk/facts)

pectedly tainted with toxic chemicals. In Belgium in 1999, for example, chicken farmers noticed signs of acute poisoning in their flocks. An investigation revealed that the potent carcinogen dioxin had somehow contaminated the chickens' feed.[12] And in Taiwan recently, 30 percent of the rice crop was found to be contaminated with arsenic, cadmium, and mercury.[13]

Food Poisoning

Proponents of the global food system would have us believe that industrial processes have left our food all but free of bacteria, but the data do not support that contention. A recent US study found one in five samples of supermarket ground meat and poultry contaminated with salmonella, while another study found the sometimes fatal germ *enterococcus faecium* in 86 percent of the supermarket chickens tested.[14]

In fact, food poisoning incidents have risen in tandem with the growth of the industrial food system. According to the Centers for Disease Control and Prevention, salmonella-related illnesses in the United States have doubled in the last two decades, and similar increases are reported for illnesses from

E. coli, campylobacter, and lysteria bacteria.[15]

In the United Kingdom, food poisoning incidents increased five-fold between 1982 and 1999, according to the British Public Health Laboratory Service (see figure 4.1). In 1997 alone over 54,000 food poisoning incidents were reported in England and Wales.[16] That sounds like a lot, but the situation may actually be far worse. Research has shown that the ratio of unreported to notified cases is 30:1—pushing the 1990s annual average to 1.4 million food poisoning cases per year.[17]

Although most cases of bacteria-tainted food are the result of unsanitary conditions in the large-scale facilities that mass-produce and process foods, the response from corporate agribusinesses and health regulatory agencies has nothing to do with cleaning up, let alone reducing the scale of, the global food system. In the United States, approval has instead been granted for irradiation as a method of sterilizing meat and other food products. Although polls indicate that more than three-fourths of the US public does not want to eat irradiated food,[18] this techno-fix is cheaper for industry and allows the fundamentally flawed global food system to go unchallenged.

In the long run, however, this solution is likely to create more problems than it solves. A large body of scientific evidence shows that irradiation reduces the nutritional value of food and leaves byproducts in the food that are themselves health hazards.[19] Although e-beam technology is now being hailed by the food industry as a safe alternative to gamma ray irradiation—which uses radioactive materials—their effects on food are the same. According to Public Citizen's Critical Mass Energy Project,

> Food irradiated by either process is deficient in vitamins and other nutrients, has caused serious health problems in laboratory animals, tastes and smells worse, is bereft of beneficial microorganisms that keep botulism and other potential deadly maladies at bay, may contain carcinogens and mysterious chemical compounds, and in the case of meat may still be tainted with faeces, urine, pus and vomit resulting from filthy slaughterhouse practices.[20]

Techno-fixes like irradiation provide, at best, temporary solutions to food safety problems whose roots lie in the excessive scale of the global food system. But proponents of global food never cease imagining that those problems can be eliminated by ratcheting up the scale one more notch. For example, Ray Goldberg, professor of Agriculture and Business at Harvard's Business School, believes that the proper response to food scares is simply to apply more technology and "to barcode every product, from a grain of cereal to a loaf of bread." And even though the increasing scale of the global food system is responsible for most food safety problems, Professor Goldberg stubbornly believes that scaling up is the solution: "These huge multinational corporations that have huge plants throughout the world have to lead the way . . . and as consolidation grows in the food system, it will become safer."[21]

Factory Farms and Human Health

There is little doubt that animals raised on small-scale, diverse farms are apt to be healthier. When allowed to range freely, particularly in organically maintained yards and pastures, they receive more exercise, their diet is more varied, and they are exposed to commensal bacteria that help exclude, and build resistance to, harmful pathogens.[22] Some organic practitioners also argue that free-ranging animals actively seek out plants with medicinal properties that can build their resistance to illness.

When livestock production is carried out on a scale that suits the global market, however, huge numbers of animals are kept in tightly confined conditions, and the potential for disease outbreaks is much higher. The important considerations of animal welfare aside, these methods lead to the rampant use of antibiotics, which poses a significant health risk not only for the livestock, but for consumers as well since antibiotic residues can remain in meat and milk. Roughly half the 25,000 tons of antibiotics produced in the United States are used in the raising of animals for human consumption.[23]

There are other reasons for concern about the overuse of antibiotics in giant livestock operations. Some 40 to 80 percent of the antibiotics used in farming are thought to be unnecessary even under factory conditions, as 80 percent of their use is as a preventative measure and for growth promotion.[24] Overuse has already rendered some drugs ineffective and may even make some strains of bacteria untreatable. According to the Public Health Laboratory Service in Britain, a new strain of salmonella that first appeared in the United Kingdom in 1990 is resistant to at least four antibiotics and now accounts for 15 percent of all salmonella food poisoning cases.[25] The newest class of antibiotics, fluoroquinolones—viewed as the last line of defense for some human infections—are already proving ineffective against some bacteria strains. An epidemiologist for the US Centers for Disease Control and Prevention says that among public health officials "there is no controversy about where antibiotic resistance in food-borne pathogens comes from": the heavy use of antibiotics in animals is to blame.[26]

The huge amounts of manure that industrial livestock farms produce also represent a human health risk. In the Cape Fear region of North Carolina, for example, factory hog farms produce ten million metric tons of waste annually, equal to that produced by forty million people. When heavy rains hit in 1999, numerous lagoons containing the manure burst. In one case, two million gallons of hog waste spilled when a lagoon ruptured at a farm that raises hogs for a subsidiary of Smithfield Foods, the largest pork producer in the United States. Such manure spills were one reason the storm left 400,000 wells in North Carolina contaminated. Health officials expressed concern that an outbreak of gastrointestinal and other diseases, such as pathogenic *E. coli*, might be caused by contaminated drinking water.[27]

Other agribusiness livestock practices are equally alarming. Monsanto has been aggressively marketing rBGH, a recombinant form of a naturally occurring hormone, for use in dairy cows. The use of the genetically engineered hormone increases milk production by 15 percent or more, but has numerous side effects: treated cows do not live as long; they are prone to develop mastitis (an infection of the udder, usually treated with antibiotics); and they often give birth to deformed and stillborn calves. As far as human health is concerned, perhaps most worrisome of all is that researchers have found elevated levels of another hormone, IGF-1, in milk from cows treated with rBGH. IGF-1 has been linked to increased likelihood of cancer in humans.[28]

Unfortunately for the general public in the United States, one of the very few countries where the use of rBGH is legal, the human health effects of this biotech product have hardly been explored. As Canadian farmer and author Brewster Kneen points out,

> The only actual testing of the drug is currently being carried out as an uncontrolled experiment on the American people, who are unknowingly consuming the milk from the drugged cows. They are unknowing because the drug's manufacturer has lobbied, litigated, and intimidated, with near-total success, to make labeling that would indicate whether or not milk comes from rBGH-treated cows virtually illegal.[29]

One of the most disturbing human health consequences of industrial livestock production is the spread of Mad Cow Disease across the species barrier from cows to humans, in the form of the deadly Creutzfeld-Jakob disease (CJD). It is generally believed that the cow variant of the disease, bovine spongiform encephalopathy (BSE), became widespread throughout the United Kingdom because of the practice of feeding the remains of dead livestock to cows—an "innovation" of large-scale agribusiness. Dead livestock were boiled down, ground up, and added to cattle feed, even though cows are naturally herbivores.

Another innovation of the global food system—the mechanical separation of meat—is thought to have played a role in spreading the disease to humans. The process extracts minute amounts of meat from bones by forcing it through a sieve under high pressure, resulting in a paste-like product—a legal ingredient in various cooked meat products—that may have included spinal cord tissue from infected cows.[30]

BSE eventually killed 175,000 cows in Britain; since the disease is believed to have a latency period of ten years, far more were undoubtedly infected. Although the British government initially insisted that there was no link between Mad Cow Disease and CJD in humans, it was later forced to reverse this stance and eventually ordered the destruction of every cow older than thirty months, some 2.5 million animals.[31]

By the end of 2001, more than one hundred people had died of CJD in the United Kingdom.[32] Like BSE in cows, however, CJD has a long latency period, and it is still unknown how high the death toll will eventually rise. The UK government's chief medical officer, Professor Liam Donaldson, admits, "We're not going to know for several years whether the size of the epidemic will be a small one, in other words in the hundreds, or a very large one, in the hundreds of thousands."[33] Meanwhile, cases of Mad Cow Disease have turned up in most other European countries and now in Asia as well.

The UK beef industry was still reeling from the impact of BSE when an outbreak of classic swine fever struck Britain in the summer of 2000, leading to the slaughter of tens of thousands of pigs. It is believed that the outbreak stems from practices all too common among industrial pig operations: transporting animals in contaminated vehicles and feeding them waste food containing infected meat.[34]

Problems like these are an inherent part of a food system that is so large that companies can increase their profits by millions of dollars simply by saving a few cents on each animal's feed, or by using chemicals or processing methods that reduce costs by a fraction of a percent.

We all want safe, healthy food, but we cannot rely on the global food system to provide it. The corporate food chain has grown so long and the distance between producers and consumers so vast that no one can really know how their food was grown, how it was processed, and how it was treated during its long travels. Only by localizing and reducing the scale of our food systems can we once again trust the food we eat.

Do Food Safety Regulations Work?

Although big business complains loudest about government regulatory red tape, many regulations would be unneeded were it not for the scale at which large producers now operate. A study by the US Centers for Disease Control and Prevention, for instance, points out that outbreaks of food-borne disease are more likely today because of the trend toward fewer, bigger food production facilities and longer distance distribution.[b]

But rather than reducing the scale of our food systems, the usual response to food safety problems is to call for techno-fixes too expensive for small producers to implement. In the United States, for example, the recent discovery of *E. coli* bacteria in some industrially produced fruit juices is likely to result in regulations requiring all juices to be pasteurized. The high cost of industrial pasteurizers would put hundreds of small producers out of business—even those for whom *E. coli* contamination is highly unlikely.[c] The European Union's demand that cheese producers install tile floors and stainless steel kitchens is having a similar impact, effectively putting an end to traditional cheese-making in Europe. In both cases, the markets of these small, local producers will be taken over by larger, more highly capitalized producers that can more easily absorb the costs of the regulations.

It is obvious to corporate executives, if not the general public, that food safety regulations favor the largest producers. The CEO of Smithfield Foods, one of the biggest meatpackers in the US, claimed that new food safety regulations would enable the company to "acquire" some smaller competitors: "a million-dollar hit [from regulators' fines] may be a disaster for them, but it's just a bump in the road for Smithfield," he admitted.[d]

Clearly, strict regulatory oversight is needed for the global food system, which depends on dangerous agricultural chemicals, antibiotics, growth hormones, genetically modified organisms, and the transport of perishable food from continent to continent. Unfortunately, corporate lobbyists and the revolving door between industry and government assure that regulations do little to guarantee the safety of global food. In the United States, for example, regulatory agencies such as the Food and Drug Administration (FDA) and the Environmental Protection Agency (EPA) effectively turned a blind eye as biotech foods spread rapidly through America's food system, even though biotech foods have never been proven safe for human consumption or for the environment. Thanks to their inaction, an estimated 60 to 70 percent of the food on the nation's supermarket shelves now contains some genetically altered ingredients.[e]

As things stand now, the global food system greatly benefits from the regulatory regime—weak as it is—that supposedly oversees it. Not only do regulations lull the public about issues like food safety, they also provide a government-sanctioned ceiling for corporate responsibility. For example, when beef produced by IBP, Inc., then the largest meatpacker in the United States, sickened members of two families, killing a five-year-old boy, a lawsuit against the corporation was dismissed by the court: since federal meat-inspection rules were followed, the company was shielded from claims of negligence.[f] (Fresh on the heels of their legal victory, IBP was forced to recall another 200 tons of ground beef that may have been contaminated with *E. coli* bacteria.[g])

Despite the clear shortcomings of government regulations, they provide a stamp of approval for even the most dubious of corporate activities. One reason that the US public was initially so complacent about bioengineered foods is that most Americans mistakenly believe their health and the safety of the environment are adequately protected by agencies like the FDA and EPA. Food corporations exploit this faith to the fullest: thus, when a watered-down treaty to limit the spread of genetically altered foods was recently passed, biotech company executives said the new treaty could actually help the industry by countering a perception that biotechnology is inadequately regulated: "I think it will give some members of the public a stronger feeling that there [are] appropriate amounts of oversight," claimed a public relations representative for Pioneer Hi-Bred International, a huge seed company.[h]

5

Food and the Economy

It is madness to fly food half way round the world—like the American raspberries on sale here at the height of our raspberry season—when British growers are going out of business.

—Hugh Raven, SAFE Alliance

SINCE THE DAWN OF INDUSTRIAL AGRICULTURE, farmers have been told to "get big or get out."[1] Many of them have, in fact, quit farming. Those who tried to get bigger are working harder than ever to make ends meet, with their profit margins continually eroded by corporate middlemen, global competition, and the rising cost of the technological treadmill. In many places, family farms have all but disappeared, their place taken by huge corporate-owned factory farms—run not by farmers, but by technicians, computers, and high-tech machinery. For farmers big and small, the global food system has been a catastrophe, destroying not only farmers' livelihoods, but also the economic health of entire communities and towns.

The local food movement offers a real chance to reverse those trends. Rather than getting big or getting out, farmers are instead helped to stay small, stay local, and prosper. Local food benefits consumers as well: rather than fattening the bottom lines of a few huge, distant corporations, people's food dollars instead help strengthen and revitalize their own communities and local economies.

Keeping Money in the Local Economy

In smaller, local systems, more of the money spent on food goes to the farmer, rather than giant corporate middlemen. Sometimes farmers can sell directly to consumers via farmers' markets, farm stands, CSAs, or other schemes, thereby eliminating middlemen and processors altogether. Even when selling through local shops and restaurants rather than directly to consumers, farmers receive much more for their production than when it is sold

ISEC/John Page

through supermarket chains, or traded as a global commodity. Equally important, the small shopkeeper's share of the price remains circulating in the local economy, adding to the community's overall economic health.

Small shops owned by independent retailers, in turn, are far more likely than supermarkets to sell locally-made products. A survey of eighty-one small shops, conducted in 1997 in East Suffolk, England, found products from 295 local producers—including large and small farmers, vegetable growers, wine producers, cheese and jam makers, village small-holders, beekeepers, and housewives making pies, soups, and cakes. One shop alone was selling more than forty local food products.[2] In many cases, owners of local restaurants actively seek out vegetables, cheeses, wines, and meats from local farms, not only because they are likely to be fresher or of higher quality, but because they add to the distinctiveness of the restaurant's menu. The Vermont Fresh Network, a program that encourages Vermont restaurants to buy from nearby growers, links more than one hundred restaurants with seventy-five local farms.[3]

Contrast this to corporate fast-food chains, which obtain food from huge monocultural farms and ship it to every part of the country, or across the world. The concept of local food is alien to such chains, as was proven by a McDonald's executive seeking approval for a new outlet in Vermont. When asked where the food they serve comes from, the executive replied, "From a distribution center in Connecticut." The absence of local foods is one reason that roughly 75 percent of the money spent at corporate franchises like McDonald's is immediately sucked out of the local economy.[4]

Nor are supermarkets—which tend to sell the same standardized products in all of their widely dispersed stores—likely to carry local products. For each supermarket to sell more than a token amount of foods produced nearby would jeopardize the structures and continual shareholder profits on which the entire global food system is based.

Keeping Jobs in the Local Economy

In an era when unemployment is a chronic problem, the job-sustaining nature of a local food system makes it highly desirable. Small farms, for example, employ far more people per acre than larger farms. This is largely because smaller farms are not as conducive to the use of large-scale equipment, and therefore use a proportionally higher amount of human labor instead. In the United Kingdom, for example, farms under 100 acres provide five times more jobs per acre than those over 500 acres.[5] The steady growth in the United Kingdom's average farm size over the last half century thus closely correlates with the loss of over 700,000 farm jobs in the same period.[6]

Moreover, wages paid to farm workers benefit local economies and communities far more than money paid for heavy equipment and the fuel to run it: while the latter is almost immediately siphoned off to corporate equipment manufacturers and oil companies, wages paid to workers remain in the local economy. Similarly, when farms are diversified, they can depend more on their own inputs and less on purchased chemical inputs. This is one reason that gross margins for organic farms average 15 percent higher than for chemical farms.[7] Again, less money leaks out of the local economy into the coffers of corporate agribusinesses.

The degree to which many industrial farmers now depend on manufactured off-farm inputs was revealed in an article in Canada's *Financial Post*, which described a large-scale farmer in southwest Manitoba:

> His own operation, like all those around him, was supported by occasional bursts of seasonal labor, comfortably cocooned inside an air-conditioned tractor. All agricultural inputs . . . including all tools, machinery, fuel, seed, fertilizer, pesticide, herbicide, not to mention all food, clothing, and shelter for him and his family, came from off-farm, mainly urban sources.[8]

Farmers like these are an agribusiness corporation's dream, but they contribute relatively little to the welfare of the local economy.

Just as small farms employ more people than do large farms, small shops provide more jobs than large supermarkets provide for the same volume of goods sold. Although large-scale retailers often boast about the number of jobs they provide, it is estimated that megamarkets destroy three jobs for every two they create.[9]

Working More, Earning Less

While more localized food systems support small farmers and help maintain the economic vitality of rural communities, the global food system has exactly the opposite effect. One of the most conspicuous features of that system is the ever-shrinking percentage of the price of food that farmers receive. In part, this is because a large number of corporate intermediaries—international traders, food processors, distributors, and supermarkets—are taking an ever-larger share. In 1910, 41 percent of US spending on food went to farmers, while 15 percent went to input suppliers and 44 percent to marketers. By 1990 the proportions had changed dramatically: input costs had risen to 24 percent and marketing to 67 percent of the cost of food. The farmer's share, meanwhile, had dropped to just 9 percent.[10] Similar changes have occurred in Europe. In Germany, for instance, about 20 percent of the food mark goes to the farmer today, down from some 75 percent in the 1950s.[11]

Farmers and their communities are being economically squeezed in other ways as well. Free trade policies are forcing farmers to compete with farmers on the other side of the world, many of whom work for a pittance. And while advances in technology have increased yields in the short run for globally traded commodities, those higher yields can also lead to glutted markets and declining prices. In any case, larger yields come at a cost to the farmer, who must spend an increasing part of every dollar earned on expensive equipment and inputs.

For farmers in financial circumstances like these, a drop in price of just a few cents or a yield decline of a fraction of a bushel per acre can mean the difference between breaking even and going further into debt. Unstable markets or unfavorable weather can easily lower returns enough to make even good farmers fail, almost overnight. Many do, and their farms are gobbled up into more highly capitalized farms, many of which are now owned by corporations.

Economist Richard Douthwaite sums up many of these trends in his description of changes in the Irish pig industry over the past twenty-five years. In the early 1970s, there were some 36,000 family farms rearing pigs in Ireland. Bacon factories were spread across the country, and about half of

www.arttoday.com

the value went back to the farm and local community. By 1996, only 700 pig farmers and six bacon factories remained, and only a fifth of the price of bacon now returns to the largely factory farms.[12]

Farming in the Third World

In the South, farmers who are still part of local economic systems are being pressured to enlist in the same global system that has been so destructive to farmers and their communities in the North. Development planners urge them to give up growing a diverse range of crops for themselves and their own communities and instead grow commodities for international markets. This inevitably means monocultural production, with all its agricultural and economic instability.

While farmers who participate in a local food system can count on feeding themselves with their own production, the industrial food system means selling on global markets and using the proceeds to buy food. A farmer in South America or Africa can easily be destroyed by events over which they have no control—a recession in Europe, perhaps, or a bigger-than-expected harvest in Asia. Meanwhile, the technological treadmill pushes out the smallest, least capitalized farmers altogether.

The result is that millions of farming families are being driven from the land every year. With no other source of income, they are forced to leave their ancestral homes and rural communities for the anonymity of one of the South's ever-expanding urban slums. Here they will likely subsist on the economic margins, constantly searching for scarce jobs. The ones who do find jobs become part of the huge pool of cheap labor, toiling in factories and sweatshops contracted out to transnational corporations. And, of course, food for these newly urban people must now be transported to reach them.

Figure 5.1: Changes in Specialist Retailing Outlets in the UK (1990–95)

Retailers	1990	1995	% Change
Butchers	17,044	15,150	–11%
Greengrocers	14,339	12,400	–14%
Bakers	6,656	5,500	–17%
Fishmongers	2,974	2,050	–31%
Total	**41,013**	**35,100**	–14.5%

Source: Retailing Enquiry, Mintel, March 1996

This adds further to pollution and resource use and increases the need to expand all the infrastructures on which the globalized system depends.

The Decline of Rural Economies

As the global food system amalgamates more and more local systems, farmers' fortunes in both the North and South have spiraled steadily downward, taking entire rural economies with them. In the North, local businesses that support the farm community have seen their customer base shrink dramatically as farmers and farm laborers are driven from the land. Thus, when 235,000 US farms failed during the mid-1980s, roughly 60,000 other rural businesses also went under.[13]

Hard times for local businesses have been compounded by the arrival of large-scale chain retailers. By 1992, the building of 25,000 out-of-town retail developments in the United Kingdom had corresponded with the closing of roughly 238,000 independent shops in villages and high streets.[14] In the 1990s alone, some 1,000 independent food shops—grocers, bakers, butchers, and fishmongers—closed each year (see figure 5.1). Family-run grocery shops now account for only 12 percent of the UK vegetable market, while the supermarket share has grown from 8 percent in 1969 to 72 percent in 1995.[15] The story has been the same in many other countries: in Italy, for instance, the arrival of superstores known as *ipermercati* has led to the demise of 370,000 small, family-run businesses—including half of the country's corner groceries—since 1991.[16]

These corporate megamarkets systematically sap the economic vitality of the communities where they set up shop. Almost nothing they sell is produced locally, and their profits are siphoned away to fatten the bottom lines of corporations that give little back to the community. Money that in a local food system would remain circulating locally over and over again is often lost to the community forever.

How Important Are Economies of Scale?

It is often argued that large-scale producers and marketers are able to displace small farms and local shops largely because economies of scale enable them to bring goods to market at lower prices. In the long run, the argument goes, lower prices mean that consumers are ultimately better off, despite the loss of local businesses.

Even setting aside the dubious assumption that people are nothing more than consumers whose overall welfare can be reduced to their purchasing capacity, this line of reasoning is fundamentally flawed. If large-scale corporate producers and marketers sell goods at lower prices than their smaller competitors, it is largely because of hidden subsidies and ignored environmental costs, both of which are ultimately paid by the consumers these trends supposedly benefit.

There are numerous ways we are forced to subsidize global food. In the United States alone, $27 billion in tax money was earmarked for farmers in 2000, most of which went to large industrial farms.[17] Other direct subsidies went to large-scale food processors and exporters: $90 million went to companies that advertise their food products abroad, helping corporations like Campbell's Soup and McDonald's take over the markets of local producers in other countries. Roughly $500 billion in public funds went to food exporters, thereby lowering the price of US commodities in the world marketplace. And hundreds of millions of dollars in loans and loan guarantees made it less costly to finance the sale of food exports.[18]

The United States is not alone in subsidizing large-scale farms and food exports; similar support for global food can be found in many other countries. But the net effect of these subsidies is to artificially lower the price of industrial food, thereby undermining local—which means *unsubsidized*—agriculture worldwide.

How much of an impact do taxpayer subsidies have? Consider the state of Montana, where 90 percent of the principal monocrop, wheat, is exported. Direct government support made up *100 percent* of the state's overall farm income in 2000.[19]

Indirect subsidies for global food are substantial as well. Consumers pay—via their tax dollars—for everything from research into chemical and biotech agriculture to the military expenses of keeping the supply of oil flowing. The public also picks up most of the tab for the costly infrastructure requirements for large-scale, globalized enterprises—including long-distance transport networks, instantaneous global communications facilities, and centralized energy infrastructures (see Chapter Two).

There is no doubt that the global food system needs these massive infrastructures. A Nebraska supermarket may sell lettuce from Mexico, apple juice made from Chilean or Hungarian apples, breakfast cereal produced in Minnesota from grain grown in Kansas, bottled water from France, and canned soup manufactured in New Jersey from ingredients produced in five other states. Similarly, farmers in Nebraska may use equipment manufactured in Japan and chemical inputs produced in California, and see their entire crop shipped to Russia.

The more localized a food system, the less need it has for global transport and communications networks. A small-scale organic farmer—one who depends largely on local labor, on inputs derived from the farm, and on markets in a thirty-five-mile radius—gets no benefit from them, nor does a small shop that sells mostly local products. In this way, subsidies for large-scale infrastructures selectively benefit the larger, global enterprises. This is a major reason why big businesses have been able to undercut—and ultimately put out of business—their smaller and more local competitors.

Foods that have been industrially produced and transported great distances also seem cheaper because they never include their environmental costs. The immense environmental toll that results from shipping foods long distances, for example, is not paid by the producer, the marketer, or the individual consumer, but by society as a whole.

Similarly, huge external costs are incurred by the industrial farming practices on which the global food system relies. According to a recent study, those costs—including damage to soil, air, water, human health, and biodiversity—amount to $3.9 billion annually in the United Kingdom, $2 billion in Germany and $34.7 billion in the United States.[20] Just removing traces of farm-related pesticides from drinking water in the United Kingdom, for instance, costs almost $200 million per year, the equivalent of $47 per hectare of chemically farmed land.[21]

One estimate of all the benefits received by US corporations from subsidies and externalized costs is $2.4 *trillion* annually.[22] Although fully quantifying ecological and social costs is impossible—no one can put a dollar value on a species made extinct, for example—it is clear that corporations have grown so large and powerful not because of their economies of scale, but because they have succeeded in passing so many of their costs on to the

More About Cheap Food

Proponents of global food claim that if the efficiencies inherent in free trade and industrial agriculture were eliminated, food prices would explode. But as this chapter has shown, global food is already expensive, although many of its costs are masked by taxpayer subsidies and externalized costs. As researcher Jules Pretty points out, "You actually pay three times for your food: once over the counter; twice through your taxes, which are used largely to support [industrial] farming; and thrice to clean up the mess caused by this method."[a]

Nonetheless, the notion that the industrialized world enjoys cheap food is a deeply ingrained myth, one that is bolstered by a superficial look at economic statistics. In the United States, for example, only 11 percent of the nation's disposable income was spent on food in 1997, compared with 22 percent in 1949[b]—numbers cited to show that food costs only half as much as it did fifty years ago. However, the social and economic changes imposed by the current development model render direct comparisons of these food expenditures largely meaningless.

For one thing, the income required to make ends meet today is much greater than it was fifty years ago. Until fairly recently, most American households had only one wage earner, and that single income was sufficient to feed, clothe, and provide housing for an entire family. (What's more, households were generally larger in the recent past, including not only more children, but often grandparents as well.)[c] Economic "progress," however, has meant that the overall expense of raising a family—even today's smaller nuclear families—increasingly requires the incomes of two wage-earners.

One of the biggest contributors to the need for ever-larger incomes is housing, an expense that continues to rise rapidly. This is a direct consequence of the same economic policy choices that have supposedly lowered the cost of food. Those policies have promoted urbanization by sucking jobs out of rural areas and centralizing them in a relative handful of cities and suburbs. In those regions, the price of land skyrockets, taking the cost of homes and rentals with it.

Thus, the proportion of income spent on food today may be less, but since total income needed is so much higher, people pay much more for food now than the statistics would lead us to believe.

There are still other reasons global food really isn't cheap. Perhaps most important of all, the economic policies that have given us industrial food have been part and parcel of a development model that involves, on a massive scale, painful dislocations of people, communities, and entire cultures. Every advance in the industrialization of agriculture has reduced the need for farmers and agricultural laborers, has gutted rural economies and communities, and has given rise to massive urban migrations. The new emphasis on trade is accelerating those effects worldwide.

Even though small farmers in the industrialized world are still being driven from the land, it is easy to believe that this process is largely over, that to be urbanized is a natural human condition. But today, 96 percent of the world's remaining agricultural workers—some 2.5 billion people—are in the South.[d] To continue promoting a model that will draw most of them into cities—consigning hundreds of millions of people to the most horrendous poverty—is evidence of little humanity and even less sense. There is no way to put a price tag on the damage that will result, but you can be sure the cost of cheap global food will not include it.

public, as well as to future generations.

Accounting for these costs makes it clear that global food is not so cheap after all. And if the direct subsidies now devoted to global food were shifted instead to locally-produced foods, then consumers would quickly find that the least costly food on market shelves is local and organic.

Are Large Industrial Farms More Productive?

Another argument often heard in support of the global food system is that industrial farming—with its consolidation of land, mechanization, and use of chemicals—has vastly increased agricultural productivity. Even those who are aware of the social and ecological devastation that results from the global food system tend to see it as a necessary evil. This view is based on the assumption that there are now too many people on the planet—many of them already malnourished—to go back to more socially and ecologically sensitive forms of agriculture.

However, if providing food for the world's hungry is the priority, then

USDA

we should begin the shift toward local food systems immediately, since they do a far better job of feeding people. The global economy treats food like any other tradable commodity, which for the sake of economic efficiency should be sold wherever the highest price can be obtained. This means poor people go hungry while surrounded by fertile land that produces luxury crops for the rich on the other side of the world. According to the Food and Agriculture Organization (FAO) of the United Nations, 61 percent of Indian children under five are malnourished,[23] yet India is one of the top food-exporters in the Third World.[24]

In any case, the superior productivity of industrial agriculture is largely a myth, one that has been propagated for years by its proponents and beneficiaries. Study after study carried out in many locations all over the world show that small-scale, diversified agricultural systems have a higher total output per unit of land than large-scale monocultures.[25] The higher productivity of small farms is all the more remarkable in light of the fact that large landholders control most of the best land, while small landholders—particularly in the South—have been pushed to more marginal plots.

The productivity of small-scale diversified farms has been consistently underestimated for several reasons. One is that government agricultural agencies tend to emphasize *labor* efficiency over *land* efficiency. Since local food systems rely much more heavily on labor-intensive rather than capital-intensive methods, productivity per unit labor is of course much lower on small

USDA/Gene Alexander

farms than on highly mechanized farms. But smaller farms are actually far more efficient if the most productive use of land is the goal—which it should be, given that the world's population must be fed on the planet's limited amount of arable land. Smaller farms are especially desirable where unemployment is high, since the majority of people freed by agricultural mechanization are generally not free at all but are merely unemployed.

Another point of confusion is between *yields* and *total output*. The term yield usually refers to the amount of any given crop produced per unit of land, whereas total output refers to the combined output of *all* products from the farm. A typical industrial farm will likely specialize in one or two products, resulting in high yields for those products compared to small, diversified farms. Comparing only the yields for both types of farm is misleading, because it ignores most of the useful products a diversified farm produces, including its own fertilizer and the ability to absorb its own waste. A study of farms in parts of West Bengal, India, found that rice fields contained 124 species of economic importance to the farmers.[26] Because official agricultural accounting would consider only the output of rice from these farms, all the other products are statistically invisible.

According to Food First's Peter Rosset, even conventional agro-economists now acknowledge an "inverse relationship between farm size and output." Smaller farms are anywhere from 200 to 1,000 percent more productive per acre than larger farms. Smaller farms in the United States, for example, produce more than ten times as much value per hectare as large farms. In Uganda, one hectare farms are four times as productive as six hectare farms, while in Syria, half hectare farms are four times as productive as eight hectare farms.[27] Similar relationships hold true in every country for which data is available.

Nonetheless, governments everywhere promote a scaling-up of farm size through policies that encourage farmers to produce monocrops for the global market. In the South in particular, policymakers believe that this will enable them to pursue the same path taken by the industrialized nations. Too often ignored, however, is that the economic power of the industrialized North has been achieved using much of the South's resources and relying on much of the South's labor. The countries of the South have no colonies to exploit, no reservoirs of resources beyond their borders to claim, nor access to a vast pool of cheap labor other than their own. The road to riches pursued by the industrial powers is now a dead-end.

Governments are also deeply wedded to the notion that limitless economic growth is synonymous with societal health. This, too, is a fallacy. A smaller-scale, diversified farmer does not need to market much production to make ends meet (and therefore adds little to the country's GDP) if draft animals are employed, if on-farm inputs like composted manure are used, and if most of the food consumed at home is grown on the farm. A farmer hooked to the industrial system, on the other hand, will need to market a much larger amount in order to pay for tractors, fuel, and chemical inputs, as well as food for the family. The industrial farmer's quality of life may actually be worse than that of the smaller, diversified farmer's, but governments remain biased toward the farm practices that add the most to GDP.

Ideas That Work
Closing the Loops

In a small town in Vermont, Carl Hammer has launched an imaginative program to transform vegetable waste from local restaurants into eggs those same restaurants need.

Rather than tossing vegetable scraps and plate scrapings into a waste bin and paying to have them trucked to a landfill, restaurants are instead paying the equivalent tipping fee to Mr. Hammer's company, Vermont Compost, which uses a mule-drawn wagon to pick up the scraps on a regular basis. The food scraps are spread out on a pile of ground hardwood bark from local sawmills. A small flock of chickens is then allowed to forage on the pile, leaving their droppings on it and eventually transforming it into a high-grade compost, much in demand by nearby farmers and gardeners. The chickens, in turn, lay eggs that are sold back to the restaurants that supplied the scraps.

The chickens need no nourishment other than the vegetable matter supplied by the restaurants. This eliminates the need to feed the hens grain, which would have to be trucked into Vermont from as far away as the Midwest or Canada.

So far, Vermont Compost has joined with two area restaurants, which supply roughly a ton of food scraps each week, enough feed for 75 hens. The hens in turn are producing roughly three dozen eggs a day. Since neither of the restaurants serves 100 percent organic food, the eggs produced cannot be certified as organic, but nonetheless are extremely high quality and command a market premium. At a price of $2 per dozen—midway between that of commercial and organic eggs—demand is already higher than the composter can meet. The operation is expected to expand steadily, bringing in more local restaurants—and more chickens—over time.

The elegance of Carl Hammer's idea is that two things of value—compost and eggs—are being created out of something that previously cost money to throw away. And since eggs formerly trucked in from elsewhere are now home-grown, more money stays in the local economy—enough, in fact, to create one full-time job so far. Pollution and fossil fuel use are reduced as well, and the compost created is adding fertility to farms and gardens that produce other local foods.

6

Food and Community

[There] isn't much community inside a big supermarket. There, we shop as isolated individuals, each in our own private world. Gone are the relationships with the soil, the grower, and, for the most part, even the distributor. Do you know the name of the produce manager in your supermarket? Or anything about his or her family?

—Art Gish, *Food We Can Live With*

IN ECONOMIES WHERE THE GAP between the richest and poorest is relatively narrow and no one lacks the necessities of life, anger and frustration are minimized, while feelings of mutual interdependence—the essence of community—are strong. Such circumstances are common in prosperous local economies, which tend to spread their gains evenly through the entire community. This is not the direction economic globalization is taking us: it is instead leading to an ever-widening gap between rich and poor, enabling some to accumulate vast fortunes while relegating others to abject poverty. In 1960, the income of the richest fifth of the global population was thirty times that of the poorest fifth; by 1997 the gap more than doubled, with the richest fifth receiving seventy-four times more than the poorest fifth.[1]

Many people claim that accelerated economic growth will solve this problem. But economic growth does not lift all boats equally: for example, the longest continuous economic expansion in America's history, running from 1992 to 2001, mainly benefited the affluent, causing the nation's rich-poor gulf to widen further.[2] If trends like these continue, social unrest and violence are sure to increase dramatically.

If the goal is to provide the most benefits to the most people, a shift toward local food would be an important first step. Improving the economic welfare of farmers, farm workers, small producers, and shopkeepers benefits entire local economies, providing in turn deep social benefits to communities as a whole. Some of those advantages would be clearly visible at first glance: instead of business districts with boarded-up storefronts or rows

ISEC/John Page

of identical corporate-owned stores, towns and villages could once again become lively, distinctive places, centers for social interaction and cultural vitality.

Webs of Interdependence

In the 1940s, an important study neatly documented the social and economic benefits of small-scale family farms to a community. Researcher Walter Goldschmidt studied two similar-sized rural towns in California: Dinuba, where numerous family farms still dotted the adjacent countryside, and Arvin, where corporate-owned industrial farms were the norm. Goldschmidt found that the town surrounded by family farms was far better off economically, supporting twice as many businesses and generating over 60 percent more retail volume than the industrial farm town. Overall, Dinuba had a higher standard of living and a narrower gap between rich and poor than the industrial-farming community.[3]

The economic advantages of retaining family farms translated into higher scores in various indicators of social vitality as well: Dinuba had more civic organizations, more newspapers, more public recreation centers and parks, even more schools and churches than Arvin. Participatory democracy was also stronger in Dinuba: there, town decisions were often made through popular vote; in Arvin, county officials imposed most decisions from above, with little input from the public.[4]

ISEC/John Page

None of this is surprising. When the web of economic links among small farmers, processors, retailers, and consumers is strong, both the economy and the sense of interdependence characteristic of real community are strengthened as well. One researcher in England described such links among seventy-five local farmers and businesses in the town of East Suffolk:

> A wholesale family butcher buys meat from about thirty local farmers. This business cuts and sells fresh meat to other outlets, cures and smokes bacon, makes sausages and cooked meats, and provides freezer packs. These products are supplied to twenty-one small shops. In addition, the family runs two butcher shops, which in turn are sourcing other produce, such as eggs, vegetables, cakes, and preserves from twenty-four local producers.[5]

Local food systems can also provide a link between people in a community who might otherwise have little or no connection at all. A vendor at one of Connecticut's sixty-three farmers' markets pointed out that such markets provide "a great chance to get out and talk to customers, to find out what they like, to explain what we're doing." Another added, "to many of the regular shoppers, the farmers have become like friends."[6]

Farmers' markets in small towns invariably become social events, and the purchase of food can even become secondary to the social interactions the market encourages. Shoppers at farmers' markets often look forward to the event, and linger to socialize with friends, neighbors, and farmers long after their purchases are made. Almost every town that hosts a farmers' market finds itself enlivened on market day.

CSAs and other forms of direct marketing similarly strengthen bonds in a community, enabling the consumer and farmer to see one another as real people. When CSA members meet at the farm on work days or festivals, the bonds among them can grow even stronger.

Compare this to the global food system, which promotes anonymity at every turn. Consumers, farmers, processors, and distributors of industrial foods rarely know one another—and may not even live within 1,000 miles of each other. Shopping in a supermarket, meanwhile, is for most people a boring and lifeless chore, certainly not something to be anticipated with joy.

In this way the industrial food system is already quite soulless, but before the collapse of the dot-com bubble it was deemed revolutionary to use the Internet to strip still more social interaction from it. Software provider SCT, for example, claimed its merger with Internet food marketer ecFood.com would allow "an industry traditionally characterized by personalized relationships between processors and suppliers to modernize."[7] Eliminating personal relationships was deemed beneficial not only for businesses but for consumers as well. Several Web-based companies enabled shoppers to purchase groceries while comfortably cocooned in front of their computers, rather than in real markets where they would encounter real people.

Internet grocery shopping is—for now at least—just one of many failed dot-com business models. In retrospect, the flaws of those businesses are all too obvious, but the massive support they received from financial institutions and the media says much about the west's unreasoning infatuation with technology.

The Death of Rural Communities

The destructive impacts of the global food system include an overall decline in rural services. As agricultural production is industrialized and rural people are uprooted, many of the social and economic institutions of villages and small towns are consolidated or transferred elsewhere, often in the name of efficiency. Four out of ten parishes in rural England today have no shop or post office, six out of ten no primary school, and three-quarters no bus service or clinic.[8] The people who remain in these communities, meanwhile, can easily lose their sense of belonging and commitment to a particular place. When people see no economic future at home for their own children, they are effectively seeing the life of their community coming to an end.

The sense of despair in many rural communities is exacerbated by a barrage of media and advertising images emphasizing the glories of modern

ISEC/John Page

life and sending the message that rural ways have no place in a future that will be, above all else, thoroughly high-tech. In the South, the overly glamorous lives depicted in films and on television lead children in particular to see their own rural ways of life as primitive and boring by contrast. Village life—already undermined by global economic forces—can easily seem an anachronistic dead-end, while location-specific social institutions and cultural practices can appear pointless and hopelessly out-of-date.

Similarly unrealistic and one-sided media messages undermine rural self-esteem in the North. People are told that milk will someday be produced in factories, using vats of genetically engineered bacteria rather than cows; that being adept at using a computer is already a prerequisite for getting *any* job; that refrigerators will read the bar codes on store-bought food, enabling an internal computer to order more as needed; that the Internet is creating a global consciousness that redefines community; and so on. Rarely, if at all, do portrayals of the future respectfully depict rural people or land-based ways of living.

Misplaced Blame

Rural self-esteem sometimes absorbs even harder blows. In many parts of the North, farms have been disappearing at record rates for well over a generation, a trend that is a direct consequence of government policies and the power of agribusiness corporations. Yet farmers who have lost their farms are implicitly told that they have no one to blame but themselves. As Wendell Berry points out, "With hundreds of farm families losing their farms every week, the economists are still saying, as they have said all along, that these people deserve to fail, that they have failed because they are the 'least efficient producers', and that the rest of us are better off for their failure."[9]

For people whose land is taken from them—in many cases land their families have lived on and worked for many generations—the sense of shame and anger can be immense. A recent article in The *New York Times* detailed the plight of a fifty-seven-year-old farmer in Iowa who lost a farm that had been in his family for more than a century:

> A week before [he] was forced to sell his cattle, he sat and cried, tears streaming down his face, making no sound, his body shuddering. "I lost all desire to do anything," he said. ". . . This is like a death, you know."[10]

Many farmers direct those emotions inward, and suicide can be the result. In the United States, in fact, suicide is now the leading cause of death among farmers, occurring at a rate three times higher than in the general population.[11] Things are no better in the United Kingdom, where farmers are taking their own lives at the rate of one per week.[12]

More and more, however, the anger is being directed outward. While many dispossessed rural people are coming to understand the broad systemic forces that are ruining local economies and entire cultures the world over, many others have been convinced that their problems can be traced to racial minorities or Catholics, to immigrants, to a vast Jewish banking conspiracy, or to a world government run by the UN and enforced by swarms of black helicopters. The mix of hopelessness and misdirected anger in America's economically ruined rural heartland is leading to increasing incidents of violence, played out in places like Ruby Ridge, Waco, and Oklahoma City. These events, and others like them, should be counted among the many externalized costs of the global economy.

Unmanageable Cities

Rural areas are not the only places that pay a heavy price for the globalization and industrialization of food: urban areas suffer as well. Throughout most of human history, rural areas and cities were mutually dependent. Now, as rural people are uprooted, cities are increasingly the repositories for those whose way of life has been destroyed.

In China, for example, the globalization of the economy and the modernization of agriculture will uproot an estimated 440 million people from rural areas in the next few decades, all of whom will be migrating to urban areas. According to China's vice minister of construction, 600 new cities will be needed by 2010 to handle the influx.[13] More often, rural people are drawn into cities that already have more people than they can accommodate, creating social and environmental problems that are largely unmanageable. Thanks to the systematic undermining of rural life, there are twenty more Third World cities with populations over ten million today than there were in 1970.[14] While this is often attributed to increases in overall population, the South's urban population explosion is far more closely linked to economic development. Cities like Karachi, Manila, and Lagos, which more than doubled in size between 1970 and 1990, grew twice as fast as overall population growth in their respective countries.[15]

People pulled into Third World urban centers almost invariably find themselves on the bottom rungs of the economic ladder. Cut off from their communities and cultural moorings, people from many differing ethnic backgrounds face ruthless competition for jobs and the basic necessities of life. With individual and cultural self-esteem already eroded by the pressure to live up to media stereotypes, the elements are in place for a dramatic increase in anger and hostility, particularly among young men. In the intensely demoralizing and competitive situation they face, differences of any kind become increasingly significant, and ethnic and racial conflict are the all but inevitable results.

In its most virulent form, conflict can escalate to the level of ethnic or religious cleansing. It can also result in acts of terrorism against the west, as happened on September 11, 2001. These horrifying events should be a reminder of the heavy cost of leveling the world's diverse multitude of social and economic systems. The attempt to create a global monoculture in the image of the west is ultimately an extreme act of violence, and we should be unsurprised when some people respond in kind.

Loss of Democracy

As economic scale grows, both producers and consumers become increasingly dependent on faceless middlemen who have no stake in the community, eroding the ability of individuals and communities to determine their own destiny. While people within smaller-scale economies are often deeply engaged in the decisions that affect them, this involvement rapidly diminishes once they are linked to a much larger-scale economic system. One of the authors observed this effect in the once-remote region of Ladakh, as development brought the area into the orbit of the global economy:

> In the decentralized village-scale economy, individuals had a real influence on the important decisions affecting them. They depended on people they knew, and on local resources they controlled themselves. Nowadays, as they are drawn ever more tightly into the socio-economic structure of India, each individual becomes just one of 800 million; as part of the global economy, one of over 6 billion. Their influence over the political and economic forces that affect them is being so reduced that they are essentially powerless.[16]

For most citizens in today's global economy, the levers of power can easily seem to be beyond the reach of all but corporate CEOs, industry lobbyists, and wealthy campaign contributors. Even worse, decisions that directly affect the livelihoods of millions of people are routinely made behind closed doors in huge corporations or in supranational institutions like the WTO. In these cases, there is not even the pretense of democracy: the decision makers are neither democratically elected, nor is the decision-making process open to public scrutiny.

For countries as a whole, losing food self-sufficiency can quickly lead to a loss of political independence. Discussing the so-called Food for Peace program, in which American food surpluses are given to countries with food shortages, former US Vice President Hubert Humphrey pointed out how the program would help further America's foreign policy objectives: "If you are looking for a way to get people to lean on you and to be dependent on you, in terms of their cooperation with you, it seems to me that food dependence would be terrific."[17] Not so terrific, however, for food-dependent nations.

According to José Bové, French sheep farmer and leader of the grassroots Confederation Paysanne, the tyranny of the global economy is clearly revealed in the US attempt to force European markets to accept beef treated with growth hormones—against the clear wishes of the majority of Europeans: "There have been three totalitarian forces in our lifetime. The totalitarianism of fascism, of communism, and now of capitalism. How can people try and tell us that we must import hormone-enhanced beef? What is that?"[18]

It is certainly not democracy. And yet proponents of globalization often speak as though the spread of the global economy and the spread of democracy were somehow inextricably linked.

Shifting course will not, of course, immediately change the undemocratic nature of many of today's entrenched institutions. But if the scale of our economies were reduced—with the distance between producers and consumers shortened, so that most regions and nations were less dependent on food imports channeled through giant transnational corporations—then the principles of participatory democracy could more easily be established. And if the heavy hand of global markets was lifted from farmers and small producers, decisions about what to grow and how to grow it would be based on the needs of people and nature, not on the impersonal demands of international finance.

7

Food Security

If enough diversity is lost, the ability of crops to adapt and evolve will have been destroyed. We will not have to wait for the last wheat plant to shrivel up and die before wheat can be considered extinct. It will become extinct when it loses the ability to evolve and when neither its genetic defenses nor our chemicals are able to protect it. And that day might come quietly even as millions of acres of wheat blanket the earth.

—Cary Fowler and Pat Mooney, Right Livelihood Award winners

IT SHOULD BE CLEAR THAT the heavy costs of the global food system make it a very poor choice indeed. But there is still another reason why support should be shifted away from globalizing, toward localizing food systems: food security. After all, even the healthiest, most nutritious food in the world will be of little help to people if their access to it is limited or threatened.

Corporate Control

One of the biggest threats to food security today stems from the increasing control a handful of corporations have over the world's food supply. In the United States, the three largest beef processors control 72 percent of the nation's beef-packing capacity. Another four companies control 84 percent of American cereal, and just two companies control 70 to 80 percent of the world's grain trade.[1] Five agribusinesses account for nearly two-thirds of the global pesticide market, almost one-quarter of the global seed market, and virtually 100 percent of the transgenic seed market.[2]

If the present pace of corporate mergers and acquisitions continues, future control over food is apt to become even more concentrated. For example, nine companies now dominate the global seed market. If mergers and takeovers already under consideration are approved, those nine companies will be amalgamated into just five.[3] Robert Fraley, co-president of

One Global Agribusiness

To see just how far globalization has proceeded in the agricultural realm, consider the Cargill corporation. A partial sampling of the company's operations includes:[a]

- processing plants for oranges in Brazil, Pakistan, and the United States;
- facilities in Chile to make fruit juice products for markets in the United States, Europe, and Japan;
- copra crushing facilities in the Philippines;
- hazelnut roasting plants in Turkey;
- corn and wheat milling plants in North America, Europe, and Latin America;
- flour mills and malting plants in the United States, India, Argentina, Belgium, Canada, China, France, Germany, the Netherlands, and Spain;
- cereal mills in the United States and United Kingdom;
- processing plants for soybeans, sunflower seeds, rapeseed, peanuts, flaxseed, corn, palm, and cottonseed in the United States, Latin America, Europe, and Asia;
- cotton gins in Tanzania, Zimbabwe, and Malawi;
- cocoa processing plants in the Netherlands and Brazil (and another under construction in the Ivory Coast);
- cattle feedlots and beef and pork processing plants in the United States;
- meatpacking plants supplying fresh and frozen boxed beef to grocery stores and wholesalers in the United States, Canada, Australia, and Honduras;
- production and processing facilities for poultry and eggs in the United States, United Kingdom, France, Honduras, and Thailand;
- plants that produce rock salt and processed salt in the United States, Australia, and the Caribbean; and
- phosphate mines and fertilizer factories for distribution in North and South America, Europe, and the Pacific Rim.

Besides its manufacturing and processing facilities, Cargill is also heavily involved in trade, maintaining a network of trading offices throughout the United States, Latin America, Europe, Africa, Asia, and the Pacific Rim. Commodities traded by Cargill include:

- edible and nonedible tallow acquired from US meatpackers and sold in Algeria, Brazil, Colombia, Egypt, Korea, Mexico, the Netherlands, Spain, and Venezuela;
- raw and white sugar traded from offices in Minneapolis, Paris, Geneva, Hong Kong, Mexico City, Sao Paulo, and Moscow;
- green coffee, sold to roasters in the United States, Europe, and Asia.
- sheet and block rubber; rice, palm and coconut oil; and cocoa products.

Cargill is also directly involved in bringing local agricultural systems into the global/industrial orbit through its agricultural consulting work for multilateral development banks, aid agencies, and governments. Cargill's agricultural consultants have so far worked in 116 countries.

ISEC/Steven Gorelick

Monsanto's agricultural division, admitted, "What you are seeing is not just a consolidation of seed companies, it's really a consolidation of the entire food chain."[4]

The Thanksgiving Day Meal, by A. V. Krebs

Every year, Americans sit down with their families to celebrate Thanksgiving, looking forward with mouth-watering anticipation to the bounty that will be spread before them.

But for most Americans, the turkey is not likely to be from Uncle Ray's farm, nor the potatoes from Aunt Jean's recipe, nor the biscuits from Mom's oven.

No, most Americans are more likely to find a Butterball brand turkey or maybe a Cook Family Foods ham on the table. There might also be some Jack Rabbit long grain rice, potatoes from Golden Valley Foods, and bread made from Peavey Grain. A nontraditionalist might even suggest putting a few Singleton butterfly shrimp on the barbecue grill, with the grill heated by Just Light charcoal briquettes.

There might also be private label pasta from the local supermarket, as well as tomatoes from Hunt's. The spices might be from Armour Dairy, with perhaps some Asian seasoning from La Choy. The salad oil might be from Wesson, the cheese from Miss Wisconsin, the canned beans from Van Camps, and the tomato or apple juice from Mott's. For dessert there might be Swiss Miss pudding, or a frozen dairy dessert from Healthy Choice, topped perhaps with some Reddi Whip.

While watching the traditional Thanksgiving Day football game on television, the family might want to dip into some Orville Reddenbacher's popcorn, putting another handful on a Budget Buy paper plate for future munching. Adults might also want to enjoy a bottle of Carlsberg beer as they watch the game.

All in all, it will be quite a testimonial to the cornucopia of food that Americans have come to take for granted in the land of Freedom of Choice.

Yet the fact is all that food, all those products, and all those brands came from just one company—ConAgra—the nation's second largest food processor and manufacturer. Like most vertically integrated agribusiness corporations, ConAgra operates across the food chain—from seedling to supermarket—reaping enormous profits at the expense of family farmers, workers, and consumers.

Each of the company's twenty-five branded foods has annual retail sales exceeding $100 million, one reason that six cents out of every American food dollar today goes to this one company. But that isn't enough to place ConAgra at the top of the corporate food heap: the nation's largest food business, Philip Morris, takes *ten* cents of every American food dollar. That's a bigger share than earned by all American farmers combined.[b]

Corporate control of food reaches all the way into the DNA of the world's staple crops. Biotechnologies, along with the intellectual property rights provisions of the WTO, are being used by corporations to claim ownership of many plant varieties developed and grown for centuries by indigenous farmers—including the foods on which most of the world's poor depend for their sustenance. Rice, maize, wheat, soybeans, and sorghum are already the subjects of almost 1,000 patents, more than two-thirds of which are held by biotech giants Aventis, Dow, Dupont, Mitsui, Monsanto, and Syngenta.[5] The US agribusiness Rice-Tee has even attempted to patent India's famous basmati rice, and Monsanto has sought a patent that would give the company exclusive rights over soy plants and seeds.[6]

Awareness is growing that corporations hold a tightening grip over our daily needs, but not everyone is concerned about this trend. A large proportion of our political and business elite see the corporation as an efficient and indispensable institution. They believe that giant TNCs are better at clothing and feeding people than any other institution in existence, and that the only real alternative to these giants would be communism, which has already proven itself unworkable. Thus, these leaders argue for the *removal* of existing restraints on corporate activity, so that they can more effectively carry on with their good work. Many employees of giant TNCs, too, have come to see the corporation as an indispensable force for good.

Many others hold a very different view. More and more people are questioning the value of these huge corporate entities and becoming skeptical of their altruistic claims. Because corporations ultimately depend upon the public to buy their products and services, they generally respond by cultivating an image of caring and concern. Expensive public relations campaigns are undertaken, including high-profile donations to charities and nonprofit organizations that help the poor, protect the environment, or provide educational opportunities for women and minorities. An example is the Monsanto corporation, infamous manufacturer of Agent Orange and other herbicides, and now a leading marketer of genetically altered seeds. Since 1999 the company has changed its motto to "Food, Health, Hope," redesigned its logo to include the color green, and published earnest manifestos about feeding the world's hungry.

Despite this classic corporate greenwash, the fact remains that even if most Monsanto employees—including the highest levels of the corporate pyramid—honestly wanted to redirect the corporation's mission toward sustainably feeding the world's hungriest, the rules of the game that govern global finance would prevent them from doing so. Those rules insist that corporate policies aim at profit-maximization and growth, and little else that isn't window dressing. If those policies mean people starve in one part of the world or another, so be it, as long as profits are healthy and stock prices rise. Since the scale of the global economy is so large, the link between a particular corporation's activities and such suffering is often not obvious, making it easier for employees to maintain a clear conscience while they focus on the bottom line. In any case, if a corporation were to veer too far in an altruistic direction, shareholder lawsuits would be likely, as would the prospects for a takeover by a more hard-nosed and profit-oriented competitor.

Feeding Corporations, Not the Hungry

A good example of corporate spin is the way biotechnology has been promoted with extravagant claims about its ability to produce more food for the world's growing population. The humanitarian hyperbole obscures the fact that such considerations have had almost no influence on the development of this technology. It is thus unsurprising that food biotechnology has been applied in ways that mainly benefit the companies that control it: Monsanto, for example, can be sure that farmers using Roundup Ready seeds will buy the company's herbicide as well, since the genetically engineered seeds cannot be treated with any other company's herbicide.

Biotech corporations also benefit because the use of these technologies forces farmers to buy new seeds year after year, rather than practicing the time-honored method of seed-saving. Monsanto even forces the farmers who plant the company's transgenic varieties to sign contracts stipulating severe penalties if they are caught saving seeds. Farmers have good reason to fear violating the contract: Monsanto employs private investigators to uncover illegal seed-saving, has a toll-free number for growers to turn in their neighbors, broadcasts the names of violators in radio ads, and has brought suit against hundreds of suspected seed-savers.[7]

None of this will be necessary if the Terminator gene is used. This patented technology protection system—created by corporate scientists in partnership with the USDA—renders the second generation of seeds sterile, thereby making seed-saving impossible.[8]

Despite the corporate propaganda, the hundreds of millions of dollars spent by biotech companies to engineer new crop varieties have been invested to reap profits—not to feed the world's poor. As Brewster Kneen points out:

> The biotech industry has no intention of feeding anyone who cannot pay well. But the hungry and deprived can be used to prey on the guilt of the affluent so the corporations can get their way with the politicians and the regulatory agencies, get new products to market, keep the industrial farmers of the north on the technology treadmill, and make their investors happy.[9]

The fact is, food biotechnology has no more humanitarian potential than the global food system of which it is part. Today, with food more tightly controlled by corporations than ever before, some 790 million people are undernourished[10]—even though more than enough food is produced to adequately feed everyone on the planet. In part, this maldistribution of food arises out of the global economy's perverse logic, in which it makes economic sense that luxury foods are grown on the best land in countries where people are starving, and then exported to countries where food is so abundant that obesity is a major problem. The profit from this trade goes not to the landless and starving poor, but to the wealthiest businesspeople in these nations.

In the South in particular, the switch from growing food for local consumption to producing for export has had severe repercussions. While threats to food security once came primarily from natural circumstances—crop failures due to drought or an unexpected frost, for instance—farmers hooked to the global food system face those same risks, plus many other purely economic ones. For example, if the market price for the commodity they grow is too low, or the price they paid for chemical inputs was too high, they may go hungry even after harvesting a bumper crop. Likewise, political instability in oil-exporting countries may limit the availability of fuels needed to operate heavy farm equipment, making planting, cultivating, or harvesting uneconomic or impossible.

For the multitude of displaced rural people crowding the Third World's urban slums, food no longer comes from the earth and their own toil, but from markets that demand cold hard cash. Endemic hunger is not uncommon. Even in rich industrialized countries, at least thirty-seven million people are unemployed, one hundred million are homeless, and millions more are underfed.[11] In the US, for example, a decade-long economic recovery was unable to lift an estimated 35 million Americans above the poverty line.[12]

For these people, even the barest minimum of food staples may be out of reach, reducing food security to public assistance programs that provide just enough to meet basic needs. Even this source of food is secure only as long as political winds do not shift, suddenly limiting public assistance or putting an end to it entirely.

The Dangers of a Homogenized Food Supply

The globalization of food poses other risks to food security. As the world's food supply becomes increasingly globalized, people everywhere are becoming dependent on the same narrow range of foods. This puts further pressure on land suitable for raising the relatively few grains, legumes, and meats that are the major commodities traded globally. At the same time, free trade and global market forces are eliminating many traditional crops from the market. According to Indian activist Vandana Shiva, prices in India for traditional oilseeds collapsed recently after international imports of soy flooded the market. In just one year, "sesame, linseed and mustard have started to disappear from the fields."[13] This situation is being repeated around the world, as local foods that served people for centuries are replaced by one or another of the global food system's cheap imports. In Mongolia—where mare's milk has always been a staple of the diet, and where there are twenty-five million milk-producing animals—shops now stock mostly European dairy products. In other places, numerous plant varieties that grow well in a particular environment and that were traditionally important additions to people's diets have been relegated to the status of weeds.

In the South, food diversity is being eroded not only by distorted market forces but by the psychological pressures that lead to a lust for the trappings of urban life—including such modern foods as packaged instant noodles, bottled soft drinks, and processed bread, flour, and rice. These foods are often considered "high class," and many people are eagerly trading in their wholesome, traditional foods for them.[14] Those healthy local foods may have supported the culture for centuries, but they are often associated with a past that the young, in particular, have been taught to dismiss.[15]

The degree to which agricultural diversity has been lost is staggering. In China, 10,000 wheat varieties were being grown in 1949; by the 1970s only 1,000 remained. In the United States, 95 percent of the cabbage, 91 percent of the field maize, 94 percent of the pea, and 81 percent of the tomato varieties have been lost. Overall, approximately 75 percent of the world's agricultural diversity has been lost in the last century, according to the UN Food and Agriculture Organization.[16]

The implications of this trend for food security are ominous. Not only

Pseudo-Diversity

It is easy for Northern consumers to believe that industrial agriculture and global trade have actually led to an increase in food diversity. A well-stocked supermarket can overwhelm with its apparent food choices: fifty different kinds of breakfast cereal; eighty feet of shelving devoted to fruit juices, soft drinks, and other beverages; six different brands of cottage cheese; ten varieties of potato chips. In the fruit and vegetable section, shoppers can find foods from around the world: pineapples from Hawaii, mangoes from South America, kiwi fruit from New Zealand, and avocados from Mexico.

Much of this apparent diversity is illusion, however, since the 80 percent of the supermarket that consists of processed foods offers little real choice. A close look at several different packages of crackers or canned soups will reveal virtually identical lists of ingredients. In many cases, the ten different brands are owned by the same food conglomerate—the only diversity is each one's distinct packaging.

In other cases, the seemingly expanded choice disguises a more disturbing contraction in the range of foods available in markets worldwide, since the few varieties that succeed as global commodities inevitably crowd out hundreds of other varieties that farmers once grew. Thus, dozens of apple varieties once may have grown within a few miles of a supermarket that today sells just three or

ISEC/John Page

four—those most favored by large growers. No matter that a Red Delicious is not as tasty as an heirloom apple variety: the Red Delicious looks and travels better.

It is likely that where mangoes, kiwi, pineapples, and other exotic fruits are native, the large number of varieties available in the past has been whittled down there as well. And though upscale supermarkets may now sell red, gold, and even blue Chilean potatoes alongside the white potatoes they were limited to a generation ago, this expanded consumer choice does not express greater overall diversity: when Chilean farmers who once grew more than 300 varieties of potato now grow only three for global markets, diversity has, in fact, dramatically declined.

It is also important to remember that people in the North stand at the very top of the global food system. While the North receives diverse foods from every part of the globe, what it sends back is often monocultural commodities like wheat and corn—and now Coca-Cola and McDonald's hamburgers—each of which displaces a wide range of indigenous products.

are fewer kinds of food being raised and eaten around the world, but diversity within the few remaining staples is being lost as well. If a monoculture field is more susceptible to devastation by pests and blight, the risks rise exponentially when much of the world's arable land is planted in virtually identical strains. In 1970, for example, 80 percent of the corn planted in the United States shared a common genetic heritage. When a maize blight struck, it quickly destroyed more than ten million acres of corn.[17]

On too many counts, the global food system is simply unsustainable. It relies heavily on fossil fuels, which are a limited, non-renewable resource that is also highly polluting. It causes extreme soil degradation, so much so that 30 percent of the world's arable cropland has been abandoned because of soil erosion in the last 40 years.[18] It is heavily dependent on large-scale irrigation, requiring huge dams that displace millions of small farmers and flood immense areas of fertile land—in some cases as much as is brought under new cultivation.[19]

Industrial agriculture is also part and parcel of a larger industrial model—based upon endlessly increasing economic growth and trade—whose wastes the earth can no longer absorb. Global warming, a direct product of that

ISEC/Steven Gorelick

system, is expected to lead to a rise in sea levels that will flood many productive, low-lying agricultural areas around the world. A one-meter rise in sea level, for example, would inundate farmland in Bangladesh that is now one hundred miles inland and would reduce that nation's rice production by 16 percent. Agriculture in fertile river deltas in China, Egypt, Indonesia, the Netherlands, and the United States would be similarly affected.[20] Global climate change may even halt or reverse ocean currents that now keep the climate temperate in northern latitudes. In the near future, many regions—including Britain, Scandinavia, and northern Germany—may be unable to support agriculture at all.[21]

Local Foods Add to Food Security

On almost every count, food security would increase if the growing reliance on the global food system were reversed. Instead of depending on distant, anonymous corporations for this most basic of needs, people could rely in greater measure on themselves or nearby farmers. Instead of being at the whim of impersonal market forces and profit-driven management decisions, people could rely on the interdependent bonds of a healthy community.

Since shifting toward the local would promote real diversity at every level, food security would be strengthened across the board. Instead of being flooded by cheap imports that make it uneconomical to grow locally distinct varieties, food that best fits local conditions would have a chance to thrive. Rather than monocultures highly susceptible to devastation by disease, pests, and weeds, farms would be more diverse, complex, and stable. Rather than identical varieties of crops planted everywhere, a wide range of varieties would be grown, limiting the potential for pandemic crop losses. And rather than increasing the rate at which greenhouse gases are being pumped into the atmosphere, the agricultural sector's contribution to those gases would begin to decrease.

This shift does not mean that communities now dependent on trade would see their markets disappear overnight, that people in higher latitudes would no longer have access to tropical fruits, or that a community whose crops fail could not expect help in the form of food from elsewhere. It simply means regaining a healthy balance between trade and local production, putting an end to the fiction that trade is always beneficial to all sides and that more trade is always better than less.

This shift also means recognizing that the bottom line is not the price of food, but its cost—to the environment, to rural livelihoods, to people's health, and to their sense of community. The global food system is very costly on all these counts, while the food security it provides is tenuous at best.

8

Shifting Direction

Going local does not mean walling off the outside world. It means nurturing local businesses which use local resources sustainably, employ local workers at decent wages, and serve primarily local consumers. It means becoming more self-sufficient, and less dependent on imports. Control moves from the boardrooms of distant corporations, and back to the community, where it belongs.

—Michael H. Shuman, *Going Local*

UNTIL RECENTLY, the global food system seemed unstoppable in its ability to displace local food systems the world over. As has been argued here, this trend has little to do with historical inevitability or evolution but is largely the consequence of government policies and the power of transnational corporations.

Today's economists and policymakers, backed up by conventional economic thinking, make no distinction between apples grown in an orchard and rubber balls manufactured in a factory. Food, they assume, is like any other commodity; society as a whole is further presumed to benefit if this commodity is produced and marketed as efficiently as possible. Because this way of thinking excludes so many externalities and hidden subsidies, it can easily seem efficient for people to consume food that poses long-term risks to human health, that was produced in ways that degrade the soil and pollute the environment, and that was needlessly packaged in layers of plastic and transported thousands of miles. Just as absurdly, conventional economic logic would have us believe that society is better off even if the scale of food production and marketing undermines national and local economies and increases the economic and political power of huge, unaccountable corporations.

While shortening the distance between producers and consumers would bring immense benefits, supporters of the global, industrial model will certainly decry such a shift, claiming that it would entail too much social and economic disruption. What this protest ignores, however, is the tremendous disruption and dislocation that our *current* direction entails. In the name of progress, family farms and rural communities throughout the world are being driven to extinction, and millions of people are being pulled off the land into sprawling, constantly expanding cities. It is absurd to speak as though a shift in direction—one that will reduce social as well as ecological breakdown—would entail too much disruption.

Fortunately, more people are beginning to recognize that a shift in direction is not only possible, but necessary. They are beginning to see for themselves the immense costs of severing food from its cultural and environmental moorings, and then treating it as a commodity subject to lawless speculative investment.

As a result, farmers, consumers, and environmentalists around the world are joining hands to demand shifts in policy away from the globalized model with its bias towards large-scale, monocultural production and corporate agribusiness, toward more localized food systems that promote smaller-scale diversified farms and healthier communities.

It is in the mutual interest of nations, regions, and local communities to increase their sovereignty and stability by ensuring that each has the capacity to provide for the basic needs of its own citizens—irrespective of the vagaries of the global market or of events on the other side of the world.

For an effective transformation to sustainable local food economies, change would need to take place at the international, national, and community levels.

International Level

With enough pressure from below, governments can be forced back to the bargaining table to renegotiate trade treaties such as NAFTA and GATT, this time with the interests of people and the environment—not corporations—at the forefront. Since challenging the hegemony of the WTO and international finance would be daunting for even the most powerful nation, a turnabout would be most likely to occur if groups of nations joined together with this purpose in mind. There is already a precedent for joint opposition to the global model: at the Montreal Biosafety Protocol meeting in February 2000, attempts by the United States to use the WTO to force the world's nations to remove all restrictions on the import of its transgenic food and crops was successfully opposed by the Like-Minded Group of Nations, consisting of 135 countries.

New rules of the game would allow the careful use of trade tariffs to regulate imports of goods that could be produced locally. Rejecting corporate-led trade does not mean that all food trade would end or that fellow citizens in other countries would be targeted. Rather, it would mean that jobs could be safeguarded and local resources defended against the excessive power of transnational corporations. The goal of tariffs and subsidies would not be to inhibit trade in foods that cannot be produced locally, but to encourage the growing of food that can.

For example, countries such as South Korea, which can be completely self-sufficient in their staple food, rice, should not be required to open their markets to rice from the United States or other producers. This is not free trade—it is *forced* trade. Far from benefiting those countries, such trade threatens the viability of farmers and their communities, while undermining their food security. Nor does this policy benefit the United States in the long run: encouraging American farmers to depend still more on exports only increases the pressures for monocultural production, with all the environmental, economic, and social problems this entails. International agreements could set standard tariffs on the import of unneeded staple foods to ensure that all nations maintain or restore their capacity to provide for their citizens' basic needs.

A reversal of the trend toward commodifying more aspects of life would also be in order. Thanks to patent rules and treaties such as GATT, corporations are claiming title to intellectual property that ranges from germ plasm in seeds cultivated by traditional people for millennia, to portions of the human genome. Living organisms and traditional resources should not be allowed to become commercial property controlled by corporations.

National Level

As we have seen, the globalization of food is being propelled by a vast array of hidden government subsidies, investments, tax breaks, and other incentives that overwhelmingly favor corporations and global trade. These would need to be reformed to ensure that prices reflect the environmental and social cost of food production and distribution, especially in the areas of transport, energy and agricultural subsidies.

Transport

Hidden transport subsidies for the global economy could be corrected by abandoning massive expansion programs and shifting this support toward a range of transport options that favor smaller, national, and local enterprises. This would have enormous benefits—from the creation of jobs, to a healthier environment, to a more equitable distribution of resources. Taxpayers could be reimbursed for their billions of tax dollars spent on these programs through the imposition of a hefty tax on heavy trucks, which inflict vastly greater damage to roads than do lighter vehicles.

The funds saved could be used to revitalize local communities and main street shopping areas. Depending on the local situation, transport money could also be spent on building bike paths, footpaths, boat and rail services, and, where appropriate, paths for animal transport. Even in the highly industrialized world, where dependence on centralized infrastructures is deeply entrenched, a move in this direction can be made. In Amsterdam, for example, steps are being taken to ban cars from the city's center, thus allowing sidewalks to be widened and more bicycle lanes to be built.

Energy

In the Third World, most people still live in small towns and rural communities and are, to a large extent, part of a local economy. In this era of rapid globalization, the most urgent challenge is to stop the tide of urbanization. Large dams, fossil-fuel plants, and other large-scale energy and transport infrastructures are geared toward the needs of urban areas and export-oriented production—thus promoting both urbanization and globalization.

Decentralized renewable energy infrastructures would help stem the urban tide by strengthening villages, smaller towns, and agricultural economies in general. Since the energy infrastructure in the South is not yet very developed, there is a realistic possibility that this path could be taken in the near future.

In both the North and the South, energy is heavily subsidized and labor is taxed. This leads to excessive mechanization and unemployment, and gives large, mechanized food manufacturers an unfair advantage over smaller, local, and national producers. Removing the energy subsidies now given to businesses while simultaneously reducing taxes on labor would help redress the imbalance. In the long run, these shifts would encourage businesses to adopt more sustainable and socially beneficial practices.

Financial support for small-scale renewable energy would help promote decentralized, sustainable energy production. This support would counter the effect of past subsidies for expensive nuclear and fossil-fuel plants. In a time of human-induced global warming, a rapid shift in this direction is urgently needed.

Agricultural Subsidies

Both hidden and direct agricultural subsidies now favor large-scale farms, industrial agribusinesses, and corporate middlemen, allowing them to lower their prices artificially and so invade local food economies, to the detriment of both farmers and consumers. Shifting government funding toward smaller-scale, diversified agriculture would help promote biodiversity, healthier soils, and fresher food.

Current subsidies include not only direct payments to farmers but funding for research and education in biotechnology and chemical- and energy-intensive monoculture. Many governments, particularly in the South, directly subsidize pesticides and chemical fertilizers as a means of encouraging large-scale agriculture for export. During the 1980s, for example, China's annual pesticide subsidies averaged some $285 million; Egypt's, $207 million; and Colombia's, $69 million.[1] Little if any support was given for smaller-scale, organic methods: the Pakistani government, for instance, devoted roughly 75 percent of its total agricultural budget to subsidizing chemical fertilizers.[2]

Governments even pay for the water used in industrial agriculture, either through direct water subsidies or by investments in massive irrigation projects. This artificially lowers the price of water and encourages the growing of water-intensive monocultures in areas that are naturally arid. The huge corporate farms in California's San Joaquin valley, for example, would be unthinkable without these publicly funded water projects. In addition, consumers, not agribusinesses, pay the cost of removing agrochemicals from drinking water—a subsidy that in the United Kingdom amounts to £119.6 million per year.[3]

Agribusinesses also receive huge tax breaks, such as the investment allowances and tax credits afforded the capital- and energy-intensive technologies on which large producers depend. On the other hand, smaller, more labor-intensive producers are disproportionately burdened by levies on labor such as income taxes, social welfare taxes, value-added taxes, and payroll taxes—once again giving large industrial farms and processors a huge advantage.

If these subsidies for the large and global were redirected toward smaller-scale, more localized producers, the shift toward more ecological and equitable food economies would get a major boost.

Local Level

In addition to these steps at the international and national level, numerous grassroots initiatives are already beginning to build local food economies. It is important to remember that isolated, scattered, small-scale efforts will not on their own achieve the desired transformation—instead we need to think in terms of institutions and structural changes that will promote "small scale on a large scale," by allowing space for more community-based economies to flourish and spread. These institutions can be started easily by

www.arttoday.com

even a small number of community members with a common vision.

A few of the thousands of grassroots initiatives taking root around the world are described below:

Buying Local

Buy-local campaigns help local businesses survive, even when they are pitted against heavily subsidized corporate competitors. These campaigns not only prevent money from leaking out of the local economy, they also help educate people about the hidden costs—to the environment and to the community—of purchasing artificially cheap, distantly produced products. Across the United States, Canada, and Europe, grassroots organizations have emerged in response to the intrusion of huge corporate marketing chains into rural and small town economies. For example, the McDonald's Corporation—which opens about five new restaurants each day[4]—has met with grassroots resistance in at least two dozen countries. In the United States, Canada and, most recently, the United Kingdom, the rapid expansion of Wal-Mart, the world's largest retailer, has spawned a whole network of

ISEC

activists working to protect jobs and the fabric of their communities from these sprawling superstores.

But as more people become aware of the advantages of buying local, structures that reduce the distance between producers and consumers are required. It is difficult, for example, for people to boycott a nearby supermarket chain when affordable alternatives for obtaining food are not available. Here are some of the strategies communities are using right now to shorten the distance between producer and consumer:

- *Farmers' markets* benefit local economies and the environment by connecting farmers directly with urban consumers. These markets offer fresher food to consumers and lower prices, while increasing the farmers' income (see box, page 20).

- Box schemes and other forms of *community supported agriculture*, whereby customers order regular boxes of in-season vegetables (and often eggs, meat, and dairy products) from local farms, can give farmers more security by providing them with a guaranteed and stable market (see box, page 23).

- *Local food co-ops* are small retail outlets that bring together local farmers, producers, and consumers seeking to revive the local food economy. These are much preferable to conventional producer co-operatives, which usually have the narrower goal of giving small farmers more leverage in the global marketplace.

Economic Structures

The above measures can help revive a local food economy. A number of economic and monetary schemes can make the task easier—by giving small businesses access to cheap loans and by ensuring that money circulates within the community rather than draining out into the pockets of giant middlemen.

- In a number of places, *community banks* and *loan funds* have been set up, thereby increasing the capital available to local residents and businesses, and allowing people to invest in their neighbors and their community rather than in distant corporations. These schemes enable small enterprises such as local farm co-ops and box schemes to obtain cheap start-up loans of the kind that banks offer only to large corporations.

- A way of guaranteeing that money stays within the local economy is through the creation of *local currencies*—alternative scrip that is used only by community members and local participating businesses. Similarly, Local Exchange Trading Systems (LETS) are, in effect, large-scale, local barter systems. People list the services or goods they have to offer and the amount they expect in return. Their account is credited for goods or services they provide to other LETS members, and they can use those credits to purchase goods or services from anyone else in the local system. Thus, even people with little or no "real" money can participate in and benefit from the circulation of credit within the local economy.

Local Food Regulations

As seen in Chapter Four (see box, page 61), food safety regulations are often inadequate to protect consumers and the environment from the hazards of the global food system. At the same time, those regulations often make it impossible for the small producer to survive.

How can regulations on large-scale operators be tightened without plac-

ISEC/Steven Gorelick

ing a killing burden on small operators? One solution to this dilemma is a two-tier system of regulations: stricter controls on large-scale producers and marketers, with strong safeguards against the revolving door between regulatory agencies and big business; and a simpler set of locally determined regulations for small-scale local enterprises. Such a system would acknowledge that communities have the right to monitor foods produced locally for local consumption, and that such enterprises involve far fewer processes likely to damage human health or the environment.

Community-based minimum standards for local production and retailing would likely vary from place to place, influenced by local conditions and community values. Community peer pressure would ensure compliance with the agreed upon standards much more effectively than current national or statewide systems, which are largely anonymous and rely upon expensive enforcement mechanisms. Local regulation would allow more flexibility, encourage more accountability, and dramatically reduce the cost of both monitoring and compliance.

These highly localized community regulations would coexist with national and international regulations for goods produced in one region and sold in another. Small-scale businesses oriented toward local markets would not be burdened by inappropriate regulations, but people and the environment would be protected from the excesses of large-scale enterprises.

People Power

Today's crisis in food and farming is giving birth to powerful alliances among those working for systemic change. Despite the claims that globalization is inevitable and irreversible, experience shows that even a relatively small amount of public pressure can greatly influence government policy. European grassroots resistance to the genetic modification of foods, for example, has made it impossible for biotech multinationals and the US government to force these foods down consumers' throats. Thanks to the public outcry, many European governments have severely restricted or even banned imports of biotech seeds and foods, even at the risk of a trade war with the United States.

Another example, mentioned earlier, is the US Department of Agriculture's retreat from its attempt to impose organic standards that reflected the interests of huge agribusinesses rather than consumers and farmers. After USDA offices were flooded with some 270,000 public objections, the department backed away from these controversial rules.

The 1999 WTO protests in Seattle, meanwhile, showed what can happen when people become aware of the social, economic, and environmental implications of globalization. Consumers, environmentalists, unionists, and farmers from North and South joined hands to demand that governments end their support for globalization. People power can shift control of our food from global corporations to local communities.

Is a Shift in Direction Possible? Lessons from Cuba

In less than a decade, a remarkable shift took place in Cuban agriculture—away from chemical-intensive monoculture for export, toward the production of diverse, organic food for local consumption. More remarkable is that this shift happened with the approval and encouragement of the Cuban government.

Until 1990, most agricultural land in Cuba was devoted to a vast monoculture of sugar cane for the world market (and after the Cuban Revolution in 1959, for the Soviet bloc countries). With the earnings from this commodity, Cuba imported the chemical inputs and petroleum needed to support its agriculture, as well as much of its food—before 1990, the country was importing an estimated 57 percent of its caloric intake.[a] However, with the collapse of its crucial Soviet market and the subsequent tightening of the US embargo,

Cuba experienced—almost overnight—an 80 percent drop in its pesticide and fertilizer imports, and more than a 50 percent decline in its food imports.[b] In the face of this crisis, Cuba responded by adopting a range of strategies to diversify production, drastically lessen dependence on chemical inputs and fossil fuels, encourage popular participation in farming, and enhance national food security and agricultural self-reliance.

Foremost among these strategies was the breakup of huge state-controlled farms into smaller worker-owned and -managed farm collectives, as well as the reorientation of agriculture toward satisfying the country's basic food needs. This entailed diversifying the crops grown as well as adopting crop rotation, intercropping, manuring, and soil conservation. In addition, the country's extensive agricultural research sector reoriented itself toward low-input ecological methods, giving rise to innovations in biofertilizing and biological pest control.[c] Most of the country's plowing returned to unmechanized means: oxen were bred to replace the tractors that could not be used due to lack of fuel. To meet the new demand for agricultural labor, the Cuban government improved rural services and provided incentives for rural peoples to remain on the land, and for urban dwellers to spend time working on farms. It also relaxed price controls and restrictions on the direct sale of produce, which led to the establishment of many vibrant farmers' markets around the country.

Local food production in this period was also spurred by the many urban gardens that sprang up in cities around the country. An Urban Agriculture Department was established in 1994 to oversee these efforts, and by 1998, there were more than 8,000 official gardens in Havana, cultivated by more than 30,000 people.[d] The department's agricultural extension agents facilitated the formation of new gardens, provided training and information services, helped people obtain needed supplies and inputs, and established connections between gardeners.[e] Leading by example, the Ministry of Agriculture replaced the front lawn at its Havana headquarters with a garden of lettuce, bananas, and beans, and many of the ministry's employees worked in the garden to provide food for the ministry's lunchroom. Even the Cuban military began producing its own food needs, and a surplus besides.[f] These urban gardens significantly reduced the burden on rural areas and reduced food transport and storage. They also increased the quality and variety of produce urban dwellers consumed.

Although Cuba made these changes out of necessity, and will no doubt abandon at least some of them when the nation is no longer cut off from the global economy, its experience is nevertheless encouraging. It illuminates the possibilities of implementing a more ecologically and socially sensitive agriculture on a national scale, even for countries that have traveled far down the path of monocultural, export-led agriculture. Cuba has shown that, with the political will, governments can successfully shift from a focus on global food to a focus on local food and can implement policies that are friendly to the needs of people, communities, and the environment.

From Global to Local

THE ARGUMENTS IN THIS BOOK, we believe, make a compelling case for shifting from global to local. Yet many people find it difficult even to imagine a shift toward more local economies: "Time has moved on," they say, "we now live in a globalized world." The assumption is that the global economy has evolved naturally, and that returning to a local emphasis is somehow to go against the grain of history.

Many misconceptions can make a shift toward the local seem impractical or utopian. An emphasis on meeting needs locally, for example, can easily be misconstrued as meaning total self-reliance on a village level, without any trade at all. But the most urgent issue today is not whether people living in cold climates have oranges or avocados, but whether their wheat, rice, or milk—in short, their basic food needs—should travel thousands of miles, when they could all be produced within fifty. The goal of localization is not to eliminate all trade, but to reduce unnecessary transport while encouraging changes that strengthen and diversify economies at the community as well as national level. The degree of diversification, the goods produced, and the amount of trade naturally would vary from region to region.

Another stumbling block is the belief that a greater degree of self-reliance in the North would undermine the economies of the Third World—that countries in the South need Northern markets in a globalized economy to lift their people out of poverty. The truth of the matter is that a shift toward smaller-scale, more local production would benefit both North *and* South while facilitating meaningful work and fuller employment everywhere. Today's globalized economy requires the South to send a large portion of its natural resources to the North as raw materials; to devote its best agricultural land to growing food, fibers, even flowers for the North; and to expend a good deal of its labor in the cheap manufacture of goods for Northern markets. Rather than further impoverishing the South, producing more ourselves would allow the South to keep more of its resources, labor, and production for itself.

ISEC/John Page

As a result of globalization, millions of people in the South are being pulled away from sure subsistence in a land-based economy into urban slums from which they have little hope of ever escaping. Diversifying and localizing economic activity—in North and South—offers the majority far better prospects.

The idea of localization also runs counter to the common belief that fast-paced urban areas are the locus of real culture, while small, local communities are isolated backwaters, relics of a past when small-mindedness and prejudice were the norm. The past is assumed to have been brutish, a time when exploitation was fierce, intolerance rampant, violence commonplace—a situation that the modern world has largely risen above. These assumptions echo the elitist or racist belief that modern people are superior—or even more highly evolved—than their underdeveloped rural counterparts.

It is not surprising that these beliefs are so widespread. The whole process of industrialization has systematically removed political and economic power from rural areas and has led to a concomitant loss of self-respect in rural populations. Globalization is accelerating this process, rapidly pushing people in small communities to the periphery, while power—and even what we call culture—is centralized somewhere else.

To see what communities are like when people retain real economic power at the local level, we would have to look back—in some cases hundreds of years—before the enclosures in England, for example, or before the colonial era in the South. While relatively little information exists about those times, the relatively isolated region of Ladakh, or Little Tibet, provides some clues about life in largely self-reliant communities. Unaffected by colonialism or, until recently, development, Ladakh's traditional community-based culture was suffused with vibrancy, joy, and a tolerance of others that was clearly connected with people's sense of self-esteem and control over their own lives. But in less than a generation, this culture has been dramatically changed by economic development.

Development has effectively dismantled the local farm-based economy; it has shifted decision-making power away from the household and village to bureaucracies in distant urban centers; it has redirected the education of children, away from a focus on local resources and needs, toward a lifestyle completely unrelated to Ladakh; and it has implicitly informed them that urban life is glamorous, exciting, and easy, and that the life of a farmer is backward and primitive. Because of these changes, there has been a loss of self-esteem, an increase in pettiness and small-minded gossip, and unprecedented levels of divisiveness and friction. If these trends continue, future impressions of village life in Ladakh may soon differ little from unfavorable stereotypes of small-town life in the West.

An equally common myth that can cloud our thinking is that there are too many people to go back to the land. It is noteworthy that a similar skepticism does not accompany the notion of *urbanizing* the world's population. What is too easily forgotten is that the majority of the world's people today—mostly in the Third World—are still on the land. Ignoring them—speaking as if people are urbanized as part of the human condition—is a very dangerous misconception, one that helps fuel the whole process of urbanization. It is thus considered utopian to suggest a ruralization of America's or Europe's population, while few questions are raised about China's plans to move 440 million people off the land and into cities in the next few decades—part of the same process that has led to unmanageable urban explosions from Bangkok and Mexico City to Bombay, Jakarta, and Lagos. In these and other huge cities, unemployment is rampant, millions are homeless or live in slums, and the social fabric is unraveling.

Even in the North, an unhealthy urbanization continues. Rural communities are being steadily dismantled, their populations pushed into the spreading suburbanized megalopolises where the vast majority of available jobs are located. In the United States, where only about 2 percent of the population still lives on the land, farms are disappearing rapidly. It is impossible to offer that model to the rest of the world, where the majority of people earn their living as farmers. But where are people saying "We are too many to move to the city"?

As this book has shown, localizing the production and marketing of food—our most basic of economic needs—is an urgent priority. When applied to farming, the global economic model is giving us food that is neither very flavorful nor nutritious, at a price that includes depleted soil, poisoned air and water, and a destabilized global climate. It is destroying rural livelihoods and hollowing out communities in both North and South. And it is enabling control over food to become dangerously concentrated within large corporations, which by their nature subordinate all other concerns to the economic bottom line. Perhaps worst of all, people everywhere are being encouraged to rely on a single model of food production—one that is dangerously lacking in diversity—thereby jeopardizing food security worldwide.

On all these counts, a shift to the local would bring immense benefits. There is, however, so much momentum toward globalization that shifting direction will not be easy. Changing government policy, for example, will require overcoming powerful vested interests and entrenched ways of looking at the world. And even though there is increasing interest in local foods, the knowledge needed to restore our farms and rural communities is being rapidly lost.

There is much work to be done. The time to start is now.

Resource Guide

There are literally thousands of groups—from think tanks and public advocacy organizations to direct action campaigners—that oppose the global food system, promote local food, or both. The following list is not meant to be complete, but is merely a sampling that demonstrates the depth and breadth of a growing movement. Eschewing false modesty, we have listed our own organization first.

International Society for Ecology and Culture (ISEC)

ISEC has done pioneering local food work since its inception in 1975. In Ladakh (India) ISEC programs have helped to keep the local food system from being dismantled by subsidized food imports, agrochemicals, hybrid seeds, and monocropping—all of which are being pushed by development "experts." During the past ten years in the United Kingdom, ISEC set up the Food Links program at the Soil Association and published *Local Harvest*, thereby helping to launch Britain's local food movement. In both the US and UK, ISEC is presenting "Local Food Roadshows" to inform community groups and local governing bodies of the many benefits of more localized food production and marketing. Beyond food and farming, ISEC's books, reports, videos, lectures, and seminars present a big-picture analysis revealing the connection between seemingly disparate social and environmental problems, thereby encouraging strategic and systemic solutions.

ISEC (UK)
Foxhole, Dartington
Devon TQ9 6EB, UK
Phone: (+44)(0) 1803 868650
Fax: (+44)(0) 1803 868651
E-mail: info@isec.org.uk
Web site: www.isec.org.uk

ISEC (USA)
PO Box 9475
Berkeley, CA 94709, USA
Phone: (+1) 510-548-4915
Fax: (+1) 510-548-4916
E-mail: isecca@igc.org

Food and Farming

Sustain

This network, comprising more than 100 UK-based public interest organizations, advocates food and farming policies that enhance the health and welfare of people and animals, improve the working and living environment and promote equity. It addresses policies from the international and national to the regional and local level. Their dozens of well-researched publications include reports on food miles, on the impact of supermarkets, on food advertising, and many others.

Sustain
94 White Lion Street
London N1 9PF, UK
Phone: (+44)(0) 20 7837 1228
Fax: (+44)(0) 20 7837 1141
E-mail: sustain@sustainweb.org
Web site: www.sustainweb.org

Institute for Food and Development Policy (Food First)

Food First is a self-described "peoples think tank" and education-for-action center. Their numerous reports, books, and videos seek to reveal the root causes of hunger and poverty around the world, and to offer bottom-up solutions that treat food as a basic human right. An important contribution of their work has been to document how much more productive small farms are than large farms.

Food First/Institute for Food and Development Policy
398 60th Street
Oakland, CA 94618, USA
Phone: (+1) 510-654-4400
Fax: (+1) 510-654-4551
E-mail: foodfirst@foodfirst.org
Web site: www.foodfirst.org

Organic Consumers Association

Formerly the Pure Food Campaign, this US public interest organization is dedicated to building a healthy, safe, and sustainable system of food production and consumption. Issues in which they are involved include free trade, biotechnology, food irradiation, and organic standards. Their activities include public education, activist networking, boycotts and protests, grassroots lobbying, media and public relations, and litigation.

The Organic Consumers Association
6101 Cliff Estate Road
Little Marais, MN 55614, USA
Phone: (+1) 218-226-4164
Fax: (+1) 218-226-4157
E-mail: ronnie@purefood.org
Web site: www.organicconsumers.org

Foundation for Local Food Initiatives

This UK-based nonprofit works largely with local and regional governing bodies, development agencies, and health authorities to promote local food economies. They offer consulting services, feasiblity studies, and training and seminars.

Foundation for Local Food Initiatives
PO Box 1234
Bristol BS99 2PG, UK
Phone: (+44)(0) 845 458 9525
E-mail: mail@localfood.org.uk
Web site: www.localfood.org.uk

Institute for Agriculture and Trade Policy (IATP)

Much of the work of this Minnesota-based NGO has focused on the impact of free trade agreements on family farms, rural communities, and ecosystems. In recent years their reports and bulletins have also examined the economic toll of genetically-engineered food crops on farmers and farm-based economies.

Institute for Agriculture and Trade Policy
2105 First Avenue South
Minneapolis, MN 55404, USA
Phone: (+1) 612-870-0453
Fax: (+1) 612-870-4846
E-mail: iatp@iatp.org
Web site: www.iatp.org

genetiX snowball

This UK-based organization created a widespread campaign of nonviolent civil disobedience against food biotechnology. They believe that governments, subservient to corporate power, have failed us, and that people themselves must take direct responsibility. One way of doing so is to uproot transgenic plants—often in small, symbolic amounts—and to encourage others to take similar action. They have published a handbook detailing the hazards of GMOs, and describing how to get involved in the campaign against them.

genetiX snowball
PO Box 13
Peace and Environment Centre
43 Gardener Street
Brighton BN1 1UN, UK
E-mail: genetixsnowball@onet.co.uk
Web site: www.fraw.org.uk/gs

Center for Food Safety (CFS)

CFS lies at the other end of the spectrum from genetiX snowball. Although both target genetically-engineered food, this Washington, DC-based advocacy group relies heavily on the legal system, seeking stricter government regulation of transgenic technologies, and testing and labeling of transgenic foods.

The Center for Food Safety
660 Pennsylvania Ave, SE, Suite 302
Washington DC 20003, USA
Phone: (+1) 202-547-9359
Fax: (+1) 202-547-9429
E-mail: office@centerforfoodsafety.org
Web site: www.centerforfoodsafety.org

GRACE Factory Farm Project

This project of the Global Resource Action Center for the Environment (GRACE) seeks to eliminate the factory farm as a mode of production in North America. It provides assistance to groups fighting the building or expansion of factory farms in their communities, or trying to close down existing factory farms. Their materials include persuasive arguments against confined animal operations, as well as useful action packs.

GRACE, Inc.
215 Lexington Avenue, Suite 1001
New York, NY 10016, USA
Phone: (+1) 212-726-9161
Fax: (+1) 212-726-9160
E-mail: dhatz@gracelinks.org
Web site: www.factoryfarm.org

Pesticide Action Network (PAN)

Do you want to know about the hazards of a given pesticide, or which US-banned pesticides are produced in the United States and exported to the Third World? PAN is the place to go. This network of over 600 NGOs, institutions, and individuals is working to replace the use of hazardous pesticides with ecologically sound alternatives. They have regional offices on every continent (except Australia, which is coodinated by its Asia and Pacific center). The North American and UK offices are listed below. For others, consult their web site (www.pan-international.org).

PAN UK
Eurolink Centre
49 Effra Road
London, SW2 1BZ, UK
Phone: (+44)(0) 20 7274 8895
Fax: (+44)(0) 20 7274 9084
E-mail: pan-uk@pan-uk.org
Web site: www.pan-uk.org

PAN North America
49 Powell St., Suite 500
San Francisco, CA 94102, USA
Phone: (+1) 415-981-1771
Fax: (+1) 415-981-1991
E-mail: panna@panna.org
Web site: www.panna.org

Broader Issues

Program on Corporations, Law and Democracy (POCLAD)

Although this organization does not focus specifically on food, farming, or agribusiness, its perspective on corporate power is essential nonetheless. POCLAD researches and analyzes the legal history of the growth of corporations—particularly how they came to be treated as "persons" under the law—and is working toward a reassertion of the public's right to apply the "death penalty" to law-breaking corporations by revoking their charters. POCLAD's work is almost exclusively focused on US law, however, limiting its relevance to anti-corporate activists in other countries.

POCLAD
PO Box 246
South Yarmouth, MA 02664-0246, USA
Phone: (+1) 508-398-1145
Fax: (+1) 508-398-1552
E-mail: people@poclad.org
Web site: www.poclad.org

Third World Network

It can be difficult for citizens of the North to see the world from the perspective of their counterparts in the South. The voluminous materials put out by Third World Network—on trade agreements, biopiracy, genetic engineering, intellectual property rights, development, and more—can help bridge that gap. Their flagship magazine, *Third World Resurgence*, is a good place to start.

Third World Network
228 Macalister Road
10400 Penang, Malaysia
Phone: (+60)(4) 2266728/2266159
Fax: (+60)(4) 2264505
E-mail: twn@igc.apc.org
Web site: www.twnside.org.sg

International Forum on Globalization (IFG)

IFG is a network of thinkers and activists from around the world. Their aim is to expose the multiple effects of economic globalization, and to reverse the globalization process by encouraging efforts to revitalize local economies and communities. Their written reports clearly lay out the threats posed by the globalized economy, while their teach-ins have been instrumental in creating the movement against corporate-led globalization (if you missed them you can order audio tapes of the proceedings).

IFG
The Thoreau Center for Sustainability
1009 General Kennedy Avenue #2
San Francisco, CA 94129, USA
Tel.: (+1) 415-561-7650
Fax: (+1) 415-561-7651
E-mail: ifg@ifg.org
Web site: www.ifg.org

Research and Information

Agribusiness Examiner

Albert Krebs's on-line newsletter provides a populist's perspective on corporate control of both the US food supply and government farm policy. Snippets gleaned from corporate press releases, industry journals, and the business news are supplemented by heartfelt editorial comment and personal testaments from farmers. A searchable web site makes this a very useful tool for researching corporate abuse within the industrial food system.

A. V. Krebs
PO Box 2201
Everett, WA 98203-0201, USA
Phone: (+1) 425-258-5345
E-mail: avkrebs@earthlink.net
Web site: www.ea1.com/CARP

Action Group on Erosion, Technology and Concentration (ETC group)

This loose-knit Canadian group provides research on erosion, biodiversity, corporate concentration, intellectual property rights, and new technologies (particularly biotechnology). Formerly known as RAFI (Rural Advancement Foundation International), the group was instrumental in publicizing the development of the Terminator gene, whose name they coined.

ETC group
478 River Avenue, Suite 200
Winnipeg, MB R3L 0C8, Canada
Phone: (+1) 204-453-5259
Fax: (+1) 204-284-7871
E-mail: etc@etcgroup.org
Web site: www.etcgroup.org

Farmers' Organizations

Via Campesina

This is an international network linking small farmers and agricultural workers across Europe, Asia, Africa, and the Americas. Its numerous regional affiliates recognize that small farmers must join hands if they are to counter the forces that threaten them all. Their web site (www.viacampesina.org) has contact information on its many member organizations. Among them is

Confédération Paysanne, the French farmers' group headed by José Bové, who became an international hero for standing up to the McDonald's corporation and his own government.

Movimiento Campesino Internacional
Tegucigalpa, MDC-Honduras
Apdo. Postal 3628
Phone: (+504) 2394679
Fax: (+504) 2359915
E-mail: viacam@gbm.hn

Small Farms Association

This UK organization is working to retain the unique character of the British countryside created by many generations of farmers. They hope to do so by promoting traditional farming methods, forging partnerships between farms and retail businesses, and working to secure grants for farmers who practice less-intensive farming methods and who maintain wildlife habitats within the farm.

Mr. P. Hosking
Ley Combe Farm
Modbury
Ivybridge, Devon PL21 OTU, UK
E-mail: philip@small-farms-association.co.uk
Web site: www.small-farms-association.co.uk

Local Food Sources

Local Harvest

In large urban areas, it is sometimes difficult to find local sources of food. In the US, the Local Harvest web site (www.localharvest.org) may be of help. Type in a zip code or locale anywhere in the US, and Local Harvest will list farmers' markets, farm stands, and CSAs nearby. This is also potentially useful for travelers, who can plan to find local sources of food wherever their journeys take them.

Biodynamic Farming and Gardening Association

So far, nothing comparable to Local Harvest exists for Canadian residents. However, the web site for the Biodynamic movement (www.biodynamics.com) includes information on Canadian CSAs, listed province by province.

National Association of Farmers' Markets

The web site of this organization (www.farmersmarkets.net) provides contact information on farmers' markets throughout the UK. They also provide support for existing markets, and help in setting up new ones.

Soil Association

Among the activities of the UK's main organic certification and promotional body is a Local Food Links program. They can put you in touch with one of the 340 box schemes already underway, or help you start a new one. They also have a list of independently produced directories of local food sources throughout the United Kingdom. Many of these directories are (appropriately enough) very local themselves, serving relatively small areas.

Soil Association
Bristol House
40-56 Victoria Street
Bristol, BS1 6BY, UK
Phone: (+44)(0) 117 929 0661
Fax: (+44)(0) 117 925 2504
E-mail: info@soilassociation.org
Web site: www.soilassociation.org.uk

Note on Measurements

Most of the data used in this book use US standard measurement units, bu in some cases where source materials use metric units, those units have bee retained. The conversions between US standard units and metric for mea surements used are as follows:

1 metric ton = 1.1 tons
1 ton = .91 metric ton

1 kilometer (km) = .62 mile
1 mile = 1.61 kilometers

1 ton-kilometer = .68 ton-mile
1 ton-mile = 1.465 ton-kilometers

1 hectare (ha) = 2.47 acres
1 acre = .4 hectare

1 kilogram (kg) = 2.21 pounds
1 pound (lb) = .45 kilogram

1 liter (l) = .26 US gallon
1 US gallon (gal) = 3.79 liters

At the time of this writing, the exchange rate between US dollars an UK pounds sterling was approximately 1.5 dollars to the pound.

Endnotes

Chapter 1: From Local to Global

Epigraph: Peter Rosset, The Case for Small Farms: An Interview with Peter Rossett. *Multinational Monitor*. July–August 2000, pp. 29–33.

1. What may be considered small can vary considerably from place to place, depending upon the ecological and social context. In some regions of the tropics, for example, small may mean less than an acre or two, while in the United States a farm could be 150 or more acres and still reasonably be considered small. The USDA defines farm size by annual farm income, rather than by area. Small farms are defined as those that have gross annual sales of $250,000 or less, which allows large differences of scale within this category, as over 60 percent of farms in the United States report gross annual sales of less than $20,000. (US Department of Agriculture. *1998 Agricultural Factbook*. Washington, D.C.: USDA, 1998).
2. Based on estimates from FAO and World Bank production and distribution data. See Jules Pretty, *Regenerating Agriculture: Policies and Practice for Sustainability and Self-Reliance* (London: Earthscan, 1995).
3. G. A. Mingay, ed. *The Agricultural Revolution, 1650–1880* (London: Adam and Charles Black, 1977); Organisation for Economic Cooperation and Development. *Agriculture and Economics*. (Paris: OECD, 1965); E. Thomas, *Introduction to Agricultural Economics* (London: Thomas Nelson and Sons, 1956); Ministry of Agriculture, Fisheries and Food. *Agriculture in the UK, 1995* (London: MAFF, HMSO, 1995).
4. In this book, the term pesticide is used generically and includes insecticides, fungicides, acaricides, nematicides, and miticides.
5. Jules Pretty, *The Living Land: Agriculture, Food and Community Regeneration in Rural Europe* (London: Earthscan, 1998).
6. Statistics Canada, Census of Agriculture, *Farm Population 1991 and 1996*, www.statcan.ca/english/censusag/apr26/can3.htm, December 28, 2001; Government of Canada Digital Collections, *The Lure of the Farm: Trends in Farm Populations in Canada*, http://collections.ic.gc.ca/potato/thennow/movement.asp, December 28, 2001.

7. Farm Count at Lowest Point since 1850: Just 1.9 million, *New York Times*, November 10, 1994; US Department of Agriculture, National Agricultural Statistics Service, *1997 Census of Agriculture*: United States Data, Table 1: Historical Highlights, p. 10.
8. John Kelly, Corporations, Agencies Get Lion's Share of Farm Subsidies, *Kansas City Star*, September 9, 2001.
9. Walston Hailed as Hero by Small Farmers, *Farmers Weekly Interactive*, January 12, 1999, www.fwi.co.uk.
10. Graham Harvey, *The Killing of the Countryside* (London: Jonathan Cape, 1996) p. 16.
11. Rosset, The Case for Small Farms.
12. Joel Dyer, *Harvest of Rage* (Boulder, Colo.: Westview Press, 1998), p. 110.
13. Stewart Smith, Farming Activities and Family Farms: Getting the Concepts Right, presented to US Congress symposium "Agricultural Industrialization and Family Farms," October 21, 1992.
14. USDA, National Agricultural Statistics Service. *Agriculture Prices* (Washington, D.C.: USDA, NASS, 1999).
15. Dyer, *Harvest of Rage*, p. 113.
16. Nicholas D. Kristof, As Life for Family Farmers Worsens, the Toughest Wither, *New York Times*, April 2, 2000, p. 1.
17. IMF Statistics Department, *International Financial Statistics*, February 1999, 53: 2. (Washington, D.C.: International Monetary Fund, 1999).
18. The study compared the Gross Domestic Product (GDP) of nations with the annual revenues of corporations. Top 100 World Economies, *The CCPA Monitor*, March 2001, p.10.
19. Ian Johnson, Tens of Millions of Peasants Are Setting Off on China's New Long March to Find Hope and Work in the City, *The Guardian* (London and Manchester), November 3, 1994, p. 13.
20. Dyer, *Harvest of Rage*, p. 70.
21. Down on the Farm: Romantic Notions of Rural Hardship Have Been Promoted by Greedy Landowners, *The Guardian* (London and Manchester), February 4, 2000.
22. Caroline Lucas, The Crazy Logic of the Continental Food Swap, *Independent on Sunday*, March 25, 2001.
23. Tim Weiner, Aid to Farmers Puts Parties in "Political Bidding Contest," *New York Times*, August 4, 1999, p. A14; Dirk Johnson, As Agriculture Struggles, Iowa Psychologist Helps His Fellow Farmers Cope, *New York Times*, May 30, 1999, National Report, p. 12.
24. C. LeQuesne, *Reforming World Trade: The Social and Environmental Priorities* (Oxford: Oxfam, 1996).
25. Dan Rademacher, The Case Against the WTO, *Terrain*, Spring 2000, p. 24.

a. Dyer, *Harvest of Rage*, pp. 113–14.

Chapter 2: The Ecology of Food Marketing

Epigraph: David Orr, *Earth in Mind* (Washington, D.C.: Island Press, 1994) p. 50.

1. Winnie Hu, This Year, Crops Battle Nature's Other Extreme, *New York Times*, July 23, 2000, p. A-21.
2. Debi Barker and Jerry Mander, *Invisible Government: The World Trade Organization: Global Government for the New Millennium?* (San Francisco: International Forum on Globalization, 1999) p. 22.
3. Stefanie Böge, *Road Transport of Goods and the Effects on the Spatial Environment* (Wuppertal, Germany: Wuppertal Institute, 1993).
4. US Census Bureau, *Transportation-Commodity Flow Survey, 1997* (Washington, D.C.: US Census Bureau, December 1999), Table 7.
5. Sustain: The alliance for better food and farming, *Food Miles — Still on the Road to Ruin?* (London: Sustain: 1999), p. 7.
6. Food and Agriculture Organization, FAOSTAT, http://apps.fao.org/page/collections?subset=agriculture, December 28, 2001. A similar pattern holds for many other commodities. In 1998, the UK imported 174,570 tons of bread, while exporting 148,710 tons; imported 21,979 tons of eggs and egg products, while exporting 30,604 tons; imported 158,294 tons of pork, while exporting 258,558 tons [Ministry of Agriculture, Fisheries and Food, Overseas Trade Data System, UK Trade Data in Food, Feed and Drink (London: MAFF, HMSO, 1999)]. See also Caroline Lucas, Stopping the Great Food Swap: Relocalising Europe's Food Supply, March 2001, www.carolinelucasmep.org.uk/publications/greatfoodswap.html.
7. Herman Daly, The Perils of Free Trade, *Scientific American* 269:5 (November 1993), p. 51.
8. J. Kooijman, Environmental Assessment of Packaging: Sense and Sensibility, *Environmental Management* 17: 5 (1993); SAFE Alliance, *Food Miles Report* (London: SAFE Alliance, 1996).
9. Anon, Do You Need All that Packaging? *Which?* (November 1993), p. 5.
10. Non-biodegradable packaging is not the only waste that goes into landfills, and we are now squandering a valuable resource in our failure to use organic waste. A quarter of London's waste (900,000 tons/year) is green and putrescible, much of it from food (T. Garnett, *Growing Food in Cities: A Report to Highlight and Promote the Benefits of Urban Agriculture in the UK* [London: SAFE Alliance and National Food Alliance, 1996]). For millennia, this organic waste has been an important resource for farmers, either as compost or as food for livestock. Now, it is a liability, which must be trucked away and dumped into dense landfills.
11. Environmental Research Foundation, Incineration News, *Rachel's Environment & Health Weekly*, no. 592, April 2, 1998.
12. Supermarkets, in fact, have actively opposed the use of refillable and returnable containers, arguing that people would be unwilling to use them—even though 84 percent of consumers in a Friends of the Earth UK survey said they would be happy

to do so. Friends of the Earth (1991), Survey of Public Attitudes to Returnable Bottles, in Hugh Raven and Tim Lang, *Off Our Trolleys? Food Retailing and the Hypermarket Economy* (London: Institute for Public Policy Research, 1995).

13. Harriet Festing, The Potential for Direct Marketing by Small Farms in the UK, *Farm Management* 9: 8, 1997, pp. 409–21.
14. US Department of Agriculture. *Farmers Market Facts* (Washington, D.C.: USDA, Agricultural Marketing Service, 2001).
15. Hugh Raven and Tim Lang, *Off Our Trolleys? Food Retailing and the Hypermarket Economy* (London: Institute for Public Policy Research, 1995).
16. Department of the Environment, Transport and the Regions, *Transport Statistics: Travel to the Shops, Personal Travel Factsheet* 6, December 1999 (www.transtat.detr.gov.uk/facts, March 22, 2000).
17. Department of the Environment, Transport and the Regions, *Transport Statistics Great Britain: 1999 Edition*, Table 1.3 (www.transtat.detr.gov.uk/facts, March 22, 2000).
18. P. Kageson, Getting the Price Right: A European Scheme for Making Transport Pay Its True Costs, in Raven and Lang, *Off Our Trolleys?*
19. D. Pearce, *Blueprint 3: Measuring Sustainable Development* (London: Earthscan, 1993).
20. Rodney E. Slater, Secretary of Transportation, letter to Al Gore, March 12, 1997, p. 2; US Department of Transportation. *Highlights of the FY 1997 Transportation Budget* (Washington D.C.: US Department of Transportation, 1997).
21. Ten Questions on TENs, European Federation for Transport and Environment, Brussels, Belgium, pp. 3–6.
22. Loan & Credit Summary (projects approved at World Bank's December 1996 board meeting.)
23. Glenn Switkes, Design Chosen for First Phase of Hidrovia, *World Rivers Review* 11: 2, June 1996.
24. *Campbell Soup Company Annual Report*, 1994, p. 6.
25. Where the Admen Are, *Newsweek*, March 14, 1994, p. 34.
26. Interagency Group on Breastfeeding Monitoring, "Cracking the Code," reported in the INFACT Canada Newsletter, Winter 1997; Baby Milk Action, the UK member of the International Baby Food Action Network, leads the boycott campaign against Nestlé. For more information, see: www.ibfan.org.
27. Cited in Alan Thein Durning, Can't Live Without It, *Worldwatch*, May–June 1993, p. 13.
28. Joel Bleifuss, Will Ronald Eat McCrow, *In These Times* 19:3, p. 13.

a. US Department of Agriculture, *National Directory of Farmers Markets* (Washington, D.C.: USDA, Agricultural Marketing Service, 2001).
b. Jenny Hey, National Association of Farmers' Markets, personal correspondence,

September 8, 2000.

c. Patricia Brooks, Dreaming of Ripe, Juicy Tomatoes With Flavor?, *New York Times*, October 3, 1999, sect. 14, p. 1.

d. Center for Integrated Agricultural Systems, *Research Brief #21: Community Supported Agriculture: Growing Food . . . and Community* (Madison, Wisc.: CIAS, University of Wisconsin, 2000).

e. Soil Association, *The Organic Food and Farming Report, 1999* (Bristol, UK: Soil Association, 1999), p. 25.

f. T. Laird (1995), cited in Richard Douthwaite, *Short Circuit: Strengthening Local Economies for Security in an Unstable World* (Devon, England: Green Books, 1996), p. 308.

g. K. Furusawa (1994), Cooperative Alternatives in Japan, in P. Conford, ed. *A Future for the Land: Organic Practice from a Global Perspective* (Bideford, England: Resurgence Books, 1994).

h. Quoted in H. Festing, The Potential for Direct Marketing by Small Farms in the UK, *Farm Management* 9:8, 1997.

i. Festing, *Farm Management.*

j. This figure and the data that immediately follows comes from US Census Bureau, *Transportation-Commodity Flow Survey, 1997*, Economic Census (Washington, D.C.: US Census Bureau, December 1999), Table 7.

k. CO_2 emissions per ton-km for various transport modes are as follows: road – .248 kg; rail – .049 kg; ship – .036 kg; air – 1.447 kg. Jules Pretty, et al. The Real Cost of the British Food Basket, Centre for Environment and Society, Department of Biological Sciences, Department of Economics, University of Essex, Colchester, UK (forthcoming).

l. US Department of Agriculture, Economic Research Service, U.S. Agricultural Trade Update, February 26, 1999; Foreign Agricultural Trade of the United States, calendar year supplements. Data given in US Census Bureau, *Statistical Abstract of the United States, 1999*, Table 1120, p. 684.

m. US Census Bureau, *Transportation-Commodity Flow Survey, 1997.*

n. Department of the Environment, Transport and the Regions. *Transport Statistics Great Britain: 1999 Edition*, Table 1.13 (www.transtat.detr.gov.uk/facts, March 22, 2000).

o. Ministry of Agriculture, Fisheries and Food, Overseas Trade Data System. *UK Trade Data in Food, Feed and Drink Including Indigeneity and Degree of Processing*, updated quarterly.

p. Angela Paxton, *The Food Miles Report: The Dangers of Long Distance Food Transport* (London: SAFE Alliance, 1994), p. 20. Similar patterns can be seen for many other commodities. In 1998, the UK imported 174,570 tons of bread, while exporting 148,710 tons; imported 21,979 tons of eggs and egg products, while exporting 30,604 tons; imported 158,294 tons of pork, while exporting 258,558 tons. Ministry of Agriculture, Fisheries and Food, Overseas Trade Data System. *UK Trade Data in Food, Feed and Drink* (London: MAFF, HMSO, last revised in July 1999).

q. Pretty et al. The Real Cost of the British Food Basket.

Chapter 3: The Ecology of Food Production

Epigraph: Wendell Berry, *The Unsettling of America* (San Francisco: Sierra Club Books, 1977).

1. Bill Duesing, Diversity, in *Living on the Earth* (East Haven, Conn.: Long River Press, 1993), p. 22.
2. D. Korneck and H. Sukopp, Rote Liste in der Bundesrepublik Deutschland ausgestorbenen, verschollenen und gefährdeten Farn-und Blütenpflanzen und ihre Auswertung für den Biotop-und Artenschutz, Bonn, 1988.
3. Kenny Ausubel, *Seeds of Change*, (San Francisco: HarperCollins, 1994), p. 87.
4. Ausubel, *Seeds of Change*, p. 86.
5. Many weeds have valuable uses in their own right, as sources of food or medicine, as attractors of beneficial insects, or as ornamentals. The word weed is usually used to refer to any unwanted plant in the agricultural field, irrespective of that plant's indigeneity, usefulness, or threat to the crop.
6. Royal Commission on Environmental Pollution, *Sustainable Use of Soil.* 19th Report of the RCEP, Cmnd 3165 (London: HMSO, 1996).
7. Richard B. Alexander and Richard A. Smith, Nitrogen (as N) fertilizer use in the US, 1945–1985, *US Geological Survey Open-file Report 90–130*, 1990.
8. It is impossible to generalize about how much of applied pesticides are lost to the environment. Some are very mobile and get into water very easily; others rapidly adsorb onto soil particles and remain bound for long periods. What is clear is that sufficient amounts of pesticides are lost from agricultural systems to cause considerable external damage. For details, see G. R. Conway and J. Pretty, *Unwelcome Harvest: Agriculture and Pollution* (London: Earthscan, 1991).
9. Derek C. G. Muir and Ross J. Norstrom, Persistent Organic Contaminants in Arctic Marine and Freshwater Ecosystems, *Arctic Research of the United States*, vol. 8, Spring 1994, pp. 136–146; Derek C. G. Muir et al., Arctic Marine Ecosystem Contamination, *The Science of the Total Environment*, vol. 122, 1992, pp. 75–134.
10. Environmental Research Foundation. Frogs, Alligators and Pesticides, *Rachel's Environment & Health Weekly*, no. 590, March 19, 1998.
11. The total nitrogen input into UK agriculture amounted to 2.37 million tons per year during the 1990s, but the output in crops and animals was only 0.65 million tons. The balance is lost to the environment. J. Pretty, *The Living Land: Agriculture, Food and Community Regeneration in Rural Europe* (London: Earthscan, 1998). Phosphate also escapes from farms to contaminate water: in the UK, an estimated 43% of phosphate in water comes from agriculture (29% from livestock and 14% from fertilizers), mostly attached to soil particles from eroded land. Royal Commis-

sion on Environmental Pollution, *Sustainable Use of Soil,* 19th report of the RCEP, Cmnd 3165 (London: HMSO, 1996); P. J. A. Withers and G. C. Jarvis, Mitigation Options for Diffuse P Loss to Water, *Soil Use and Management*, vol. 14, pp. 186–92; Environmental Agency, *Aquatic Eutrophication in England and Wales: A Proposed Management Strategy* (Bristol: Environmental Agency, 1998).

12. Louisiana Universities Marine Consortium, press release, July 29, 1999; Ronald C. Antweiler, Donald A. Goolsby, and Howard E. Taylor, Nutrients in the Mississippi River, in Robert H. Meade, ed. *Contaminants in the Mississippi River, 1987–92*, US Geological Survey Circular 1133 (Reston, Va.: USGS, 1995). The dead zone is basically a huge marine monoculture. Fertilizing nutrients in the agricultural run-off dramatically increase algal growth near the Mississippi and Louisiana gulf coasts. This algae depletes the oxygen in the water (eutrophication), making it virtually impossible for fish and other marine life to survive. This process is not unique to the Gulf of Mexico—it exists near the mouths of many rivers that flow through agricultural regions where industrial methods are used.
13. David Weir, *The Bhopal Syndrome: Pesticides, Environment, and Health* (San Francisco: Sierra Club Books, 1987); Brojendra Nath Banerjee, *Environmental Pollution and Bhopal Killings* (Delhi: Gian Publishing House, 1987).
14. Helena Norberg-Hodge, *Ancient Futures: Learning from Ladakh* (San Francisco: Sierra Club Books, 1991 and London: Rider Books, revised ed. 2000). Also see Edward Goldsmith, Learning to Live With Nature: The Lessons of Traditional Irrigation, *The Ecologist* 28:3, May/June 1998.
15. Aleta Brown, Tehri Stalled by Powerful Fast, *World Rivers Review* 11:3, July 1996.
16. F. Ghassemi, A. J. Jakeman and H. A. Nix, *Salinisation of Land and Water Resources* (Wallingford Oxon, England: CAB International, 1995), p. 48.
17. Peter Goering, Helena Norberg-Hodge, and John Page, *From the Ground Up: Rethinking Industrial Agriculture* (London: Zed Books, 1993), p. 28.
18. J. E. T. Moen and W. J. K. Brugman, Soil Protection Programmes and Strategies in Other Community Member States: Examples from the Netherlands, in H. Barth and P. L'Hermite, eds., *Scientific Basis for Soil Protection in the European Community*, Proceedings of EC Symposium, Berlin, October 1986, pp. 429–36.
19. Berry, *The Unsettling of America*, p. 62.
20. Wendell Berry, What Are People For?, in *What Are People For? Essays by Wendell Berry* (San Francisco: Northpoint Press, 1990), p. 124.
21. V. H. Heywood, executive ed., *Global Biodiversity Assessment*, published for the United Nations Environment Programme by Cambridge University Press, 1995, p. 595.
22. Heywood, *Global Biodiversity Assessment*, p. 742.
23. Mike Toner, Cultivating Designer Fish, *Atlanta Journal*, May 21, 1999; Colin Tudge, *The Engineer in the Garden* (London: Jonathan Cape, 1993); Luke Anderson, *Genetic Engineering, Food and Our Environment* (Devon, England: Green Books, 1999).

24. Although the whole point of Roundup Ready seeds is that more potent doses of the herbicide can be used without endangering the marketable crop, Monsanto has attempted to confuse the public by claiming that Roundup Ready seeds *reduce* the amount of herbicide needed. While it is true that the volume of Roundup sprayed onto crops may be reduced, its potency has been increased. Before Roundup Ready seeds hit the market, Monsanto had to lobby the US Environmental Protection Agency to raise the tolerance level for glyphosate (the key toxic ingredient in Roundup) from 6 parts per million to 20, because the new line of Roundup Ready products would not be of much value if the herbicide could not be made more potent. The EPA readily consented. Marc Lappé and Britt Bailey, *Against the Grain: Biotechnology and the Corporate Takeover of Your Food* (Monroe, Maine: Common Courage Press, 1998), pp. 75–76.
25. John E. Losey, Linda S. Rayor, and Maureen E. Carter, Transgenic Pollen Harms Monarch Larvae, *Nature* 399: 673, May 20, 1999, p. 214.
26. David Barboza, Will Agbiotech's Genetic Contamination Conquer the World?, *New York Times*, June 10, 2001.
27. Sarah Yang, Transgenic DNA discovered in native Mexican corn, according to a new study by UC Berkeley researchers, press release, University of California – Berkeley, November 29, 2001.
28. Barboza, Will Agbiotech's Genetic Contamination Conquer the World?
29. Claire Miller, cited in GM Pollution Now Pervasive: Agency, www.theage.com.au, April 30, 2001.
30. Barboza, Will Agbiotech's Genetic Contamination Conquer the World?
31. Jeffrey Benner, Frankenbugs in the Wings, *Wired.com News*, December 7, 2001, www.wired.com/news/print/0,1294,48774,00.html.

a. Soil Association, *The Organic Food and Farming Report, 1999* (Bristol, UK: The Soil Association, 1999), pp. 22–24.
b. D. Cobb, R. Feber, A. Hopkins and L. Stockdale, *Organic Farming Study*, Global Environmental Change Programme, Briefing 17 (Falmer, England: University of Sussex, 1998); see also Soil Association, *The Biodiversity Benefits of Organic Farming* (Bristol, UK: Soil Association, 2000).
c. N. Lampkin and C. Arden-Clarke, Economics and the Environment—Can Organic Farming Marry the Two?, in M. Lampkin, ed., *Collected Papers on Organic Farming* (Aberystwyth, Wales: Centre for Organic Husbandry and Agroecology, 1990).
d. N. Lampkin, Welsh Institute of Rural Studies, University of Wales, Aberystwyth, UK; complete data and charts are given on the Institute's website: www.wirs.aber.ac.uk/research/.

Chapter 4: Food and Health

Epigraph: Vicki Williams, *USA Today*, July 26, 1985, cited in A.V. Krebs, *The*

Agribusiness Examiner, no. 48, September 28, 1999.

1. R. D. Morgan, ed., *Pesticides, Chemicals and Health* (London: Edwards Arnold, 1992).
2. Environmental Research Foundation, Pesticides and Aggression, *Rachel's Environment and Health Weekly*, no. 648, April 29, 1999.
3. Blue-baby syndrome is also known as methaemoglobinaemia. The condition is strongly associated with bacterial contamination in water and a range of other factors, of which nitrates in water is just one. G.R. Conway and J. Pretty, *Unwelcome Harvest: Agriculture and Pollution* (London: Earthscan, 1991); Peter Goering,, Helena Norberg-Hodge, and John Page, *From the Ground Up: Rethinking Industrial Agriculture* (London: Zed Books, 1993), p. 15.
4. Reported in S. Postel, Controlling Toxic Chemicals, *State of the World, 1988* (New York: W.W. Norton, 1988).
5. London Food Commission, *Food Adulteration and How to Beat It* (London: Unwin, 1998).
6. Environmental Research Foundation, Pesticides and Aggression.
7. Worrisome Level of Pesticide Found in Environment, *Los Angeles Times*, October 29, 1999.
8. Steven F. Arnold et al., Synergistic Activation of Estrogen Receptor with Combinations of Environmental Chemicals, *Science*, vol. 272, June 7, 1996, pp. 1489–92.
9. Jennifer Ferrara et al., *Vermont's Atrazine Addiction* (Walden, Vt.: Food and Water, Inc., 1997).
10. Environmental Protection Agency, Office of Pesticides Programs. See list of carcinogenic pesticides and their regulatory status at the EPA website: www.epa.gov/pesticides/carlist/.
11. Lennart Hardell and Mikael Eriksson, A Case-Control Study of Non-Hodgkin Lymphoma and Exposure to Pesticides, *Cancer* 85: 6, March 15, 1999, pp. 1353–60.
12. EU Gives Order to Destroy Belgian "chicken a la dioxin," *Washington Post*, June 3, 1999; Miriam Jacobs, The Belgian Dioxin Crisis, 1999, *The Ecologist* 29:6, October 1999.
13. Edward Goldsmith, Global Trade and the Environment, in J. Mander and E. Goldsmith, eds. *The Case Against the Global Economy* (San Francisco: Sierra Club Books, 1996).
14. Janet Raloff, Retail Meats Host Drug-resistant Bacteria, *Science News Online*, October 20, 2001, www.sciencenews.org/20011020/.
15. Robert V. Tauxe, Emerging Foodborne Diseases: An Evolving Public Health Challenge, *Emerging Infectious Diseases* 3:4, October–December 1997.
16. Public Health Laboratory Service (UK), facts and figures: www.phls.co.uk/facts/NOIDS/foodAnnNots.htm, December 28, 2001.
17. Unpublished June 1998 report by British physicians to the Ministry of Agriculture, Fisheries and Food (MAFF), in Jules Pretty, et al., *An Assessment of the External*

Costs of UK Agriculture (London: Agricultural Systems, 1999).

18. A. V. Krebs, *The Agribusiness Examiner*, no. 68, March 27, 2000.
19. David R. Murray, *Biology of Food Irradiation* (Taunton, UK: Research Studies Press, 1990).
20. Public Citizen, Critical Mass Energy Project, Irradiation By Any Other Name, www.citizen.org/cmep/, April 19, 2000.
21. Under the Food Scares, a Credibility Problem, *New York Times*, July 4, 1999.
22. Brewster Kneen, *Farmageddon: Food and the Culture of Biotechnology* (Gabriola Island, BC: New Society Publishers, 1999).
23. Environmental Research Foundation, Hidden Costs of Animal Factories, *Rachel's Environment & Health Weekly*, no. 690, March 9, 2000.
24. P. F. Harrison and J. Lederberg, eds., *Antimicrobial Resistance: Issues and Options* (Washington, D.C.: National Academy Press, 1998); R. Wise, et al. Antimicrobial Resistance, *British Medical Journal*, vol. 317, pp. 609–10.
25. Public Health Laboratory Service (UK), facts and figures, www.phls.co.uk/facts/.
26. Environmental Research Foundation, Hidden Costs of Animal Factories.
27. A. V. Krebs, Hurricane Floyd and Corporate Hogs: North Carolina's 18,000-Square mile Cesspool, *Agribusiness Examiner*, no. 48, September 28, 1999.
28. Environmental Research Foundation. Milk, rBGH and Cancer, *Rachel's Environment & Health Weekly*, no. 593, April 9, 1998.
29. Kneen, *Farmageddon: Food and the Culture of Biotechnology*, p. 64.
30. Centers for Disease Control, Bovine Spongiform Encephalopathy and Variant Creutzfeldt-Jakob Disease: Background, Evolution, and Current Concerns, *Emerging Infectious Diseases* 7:1, Jan.–Feb. 2001, www.cdc.gov/ncidod/eid/vol7no1/brown.htm; USDA, Meat and Poultry Labeling Terms, www.fsis.usda.gov/OA/pubs/lablterm.htm, December 28, 2001.
31. Environmental Research Foundation, Mad Cow Disease in Humans, *Rachel's Environment & Health Weekly*, no. 683, January 20, 2000.
32. Timeline: The Rise and Rise of BSE, *New Scientist*, www.newscientist.com/hottopics/bse/bsetimeline.jsp, December 28, 2001.
33. Environmental Research Foundation, Mad Cow Disease in Humans.
34. Simon Bowers and Paul Brown, Swine Fever Crisis Hits Pig Farms, *The Guardian* (London and Manchester), August 12, 2000, p. 2.

a. Adapted from World in a Grain of Rice, *The Ecologist* 30:9, December 2000/January 2001.

b. Report: Mass Production Promotes Food Poisoning, *Burlington Free Press*, December 10, 1997, p. 1.

c. Ross Hume Hall, *The Other Cider the Farm Gate*, *The Ecologist* 30: 4, June 2000, p. 31.

d. Joseph W. Luter III, cited in A. V. Krebs, *Agribusiness Examiner*, no. 125, September

19, 2001.

e. The GMO Conundrum, *Acres USA*, June 2000, p. 3.
f. A. V. Krebs, *Agribusiness Examiner*, no. 77, June 6, 2000.
g. A. V. Krebs, *Agribusiness Examiner*, no. 80, June 28, 2000.
h. Andrew Pollack, 130 Nations Agree on Safety Rules for Biotech Food, *New York Times*, January 30, 2000, p. 1.

Chapter 5: Food and the Economy

Epigraph: Cited in Richard Douthwaite, *Short Circuit: Strengthening Local Economies for Security in an Unstable World* (Dartington, Devon: Green Books, 1996), p. 282.

1. Comment made by Earl Butz, Secretary of Agriculture under Richard Nixon.
2. C. Cranbrook, *The Rural Economy and Supermarkets* (Suffolk, England: Great Gelmham, 1997).
3. Vermont Fresh Network website, www.vermontfresh.net, December 28, 2001.
4. Cited in Kirkpatrick Sale, *Human Scale* (New York: Coward, McCann and Geoghegan, 1980), pp. 88–89.
5. H. Raven and M. Brownbridge, Why Small Farmers?, cited in S. P. Carruthers and F. A. Miller, eds., *Crisis on the Family Farm: Ethics or Economics?*, CAS Paper 28 (Reading, England: CAS, 1996).
6. J. Pretty, *The Living Land: Agriculture, Food and Community Regeneration in Rural Europe* (London: Earthscan, 1998).
7. D. Cobb, R. Feber, A. Hopkins, and L. Stockdale, *Organic Farming Study*, Global Environmental Change Programme, Briefing 17 (Falmer, England: University of Sussex, 1998).
8. Canada's Farm Crisis? What Farm Crisis?, *Financial Post* (Canada), January 18, 2000.
9. Kai Mander and Alex Boston, Wal-Mart: Global Retailer, in J. Mander and E. Goldsmith, eds., *The Case Against the Global Economy* (San Francisco: Sierra Club Books, 1996).
10. Stewart Smith, Farming Activities and Family Farms: Getting the Concepts Right, presented to US Congress symposium Agricultural Industrialization and Family Farms, October 21, 1992, p. 3.
11. Greenpeace, *Green Fields, Green Future* (London: Greenpeace Publications, 1992).
12. Reported in Richard Douthwaite, *Short Circuit: Strengthening Local Economies for Security in an Unstable World* (Devon, UK: Green Books, 1996), p. 283.
13. Farm Count at Lowest Point since 1850: Just 1.9 Million, *New York Times*, November 10, 1994.
14. DoE/MAFF, *Rural England: A Nation Committed to a Living Countryside*, The Rural White Paper (London: HMSO, 1995); *TEST. Trouble in Store: Retail Locational Policy in Britain and Germany* (London: Transport and Environmental Studies, 1988).

15. D. Hughes, Dancing with an Elephant: Building Partnerships with Multiples, paper presented at The Vegetable Challenge conference, London, May 21, 1996, The Guild of Food Writers.
16. Vania Grandi, Small Grocers Disappearing into History as Superstores Emerge in Italy, *Burlington Free Press* (Vermont), January 2, 1998, p. 6B.
17. John Kelly, Corporations, Agencies Get Lion's Share of Farm Subsidies, *Kansas City Star*, September 9, 2001.
18. USDA, Agricultural Export Assistance Update: Quarterly Report, June 2001, www.fas.usda.gov/excredits/quarterly/2001/june-sum.html#market, December 28, 2001; David E. Rosenbaum, Corporate Welfare's New Enemies, *New York Times*, February 2, 1997; Leslie Wayne, Spreading Global Risk to American Taxpayers, *New York Times*, September 20, 1998, sect. 3, p. 1.
19. Timothy Egan, Failing Farmers Learn to Profit from Federal Aid, *New York Times*, December 24, 2000.
20. J. Pretty, et al., The Real Cost of the British Food Basket, Centre for Environment and Society, Department of Biological Sciences, Department of Economics, University of Essex, Colchester, UK (forthcoming).
21. J. Pretty, et al., *An Assessment of the External Costs of UK Agriculture* (London: Agricultural Systems, 1999).
22. Ralph Estes, *Tyranny of the Bottom Line: Why Corporations Make Good People Do Bad Things* (San Francisco: Berrett-Koehler, 1996).
23. Food and Agriculture Organization, *Mapping Undernutrition: An Ongoing Process* (Rome: FAO, 1996).
24. Francis Moore Lappé et al., *World Hunger: Twelve Myths* [2nd ed.] (New York: Grove Press, 1998), p. 9.
25. Peter Rosset, The Multiple Functions and Benefits of Small Farm Agriculture in the Context of Global Trade Negotiations, *Policy Brief No. 4* (Oakland, Calif.: Institute for Food and Development Policy, 1999); R. Albert Berry and William R. Cline, *Agrarian Structure and Productivity in Developing Countries* (Baltimore: Johns Hopkins University Press, 1979); Gershon Feder, The Relationship between Farm Size and Farm Productivity, *Journal of Development Economics*, vol. 18, 1985, pp. 297–313.
26. P. Howard-Borjas, Women, Environment, and Sustainable Development, an FAO issues paper for internal discussion, May 30, 1992, p. 7.
27. Rosset, The Multiple Functions and Benefits of Small Farm Agriculture in the Context of Global Trade Negotiations.

a. J. Pretty, et al., An Assessment of the Total External Costs of UK Agriculture, *Agricultural Systems* 65:2, 2000, pp. 113–36.
b. Howard Elitzak, *Food Cost Review, 1950–97*, Agricultural Economic Report No. 780, US Department of Agriculture.

c. Donald J. Hernandez, *Population Change and the Family Environment of Children*, U.S. Bureau of the Census.
d. Food and Agriculture Organization. Gender and Food Security Statistics, www.fao.org/Gender/en/stats-e.htm, December 28, 2001.

Chapter 6: Food and Community

Epigraph: Art Gish, Food We Can Live With, *The Plain Reader* (New York: Ballantine Books, 1998), p. 83.

1. UNDP, The Facts of Global Life, press kit for *Human Development Report 1999*, www.undp.org/hdro/E5.html, December 28, 2001.
2. By 2000, the portion of total income received by the wealthiest 20 percent of Americans had risen to 49.7 percent (from 46.6 percent in 1990), while the portion received by the poorest 20 percent sank to 3.6 percent (from 3.9 percent in 1990). US Census Bureau, Selected Measures of US Household Income Dispersion: 1967–2000, in *Current Population Reports, P60-213, Money Income in the United States: 2000* (Washington, D.C.: US Government Printing Office, 2001).
3. Cited in Joel Dyer, *Harvest of Rage* (Boulder, Colo.: Westview Press, 1998), p. 120.
4. Dyer, *Harvest of Rage.*
5. C. Cranbrook, *The Rural Economy and Supermarkets* (Suffolk, England: Great Gelmham, 1997).
6. Patricia Brooks, Dreaming of Ripe, Juicy Tomatoes With Flavor?, *New York Times*, October 3, 1999, sect. 14, p. 1.
7. Internet Software Provider Forms Food Alliance, *Financial Times* (London), February 17, 2000.
8. Countryside Agency, *State of the Countryside* (Cheltenham, England: Countryside Agency, 1999).
9. Wendell Berry, What Are People For?, in *What Are People For?: Essays by Wendell Berry* (San Francisco: Northpoint Press, 1990), p. 123.
10. Dirk Johnson, As Agriculture Struggles, Iowa Psychologist Helps His Fellow Farmers Cope, *New York Times*, May 30, 1999 (National Report Pages), p. 12.
11. Dyer, *Harvest of Rage*, p. 33.
12. Blair's Views on Rural Life Slated by Advisers, *The Scotsman*, February 4, 2000.
13. Ian Johnson, Tens of Millions of Peasants Are Setting Off on China's New Long March to Find Hope and Work in the City, *The Guardian* (London), November 3, 1994, p. 16; Preserving Global Cropland, *State of the World, 1997* (New York: W. W. Norton, 1997), p. 48.
14. David Morris, Unmanageable Megacities, *Utne Reader*, September–October 1994, p. 80.
15. *World Urbanization Prospects: The 1992 Revision* (New York: United Nations, 1993), table A.11.
16. Helena Norberg-Hodge, Ancient Futures, in D. Chiras, ed., *Voices for the Earth:*

Vital Ideas from America's Best Environmental Books (Boulder, Colo.: Johnson Books, 1995).

17. Cited in A. V. Krebs, Cuba Readies for Second "Bay of Pigs" Landing, *Agribusiness Examiner*, no. 134, November 26, 2001.
18. Suzanne Daley, French See a Hero in War on "McDomination," *New York Times*, October 12, 1999, p. A1.

Chapter 7: Food Security

Epigraph: Cited in Richard Douthwaite, *Short Circuit: Strengthening Local Economies for Security in an Unstable World* (Dartington, Devon: Green Books, 1996), p. 271.

1. The three largest beef packing companies are Tyson, ConAgra, and Excel (a subsidiary of Cargill); the four largest cereal companies are Kelloggs, General Mills, Philip Morris, and Quaker Oats; the companies that dominate the world's grain trade are Cargill and Archer Daniels Midland. Updated View of the Meatpacking Industry, *REAP News and Views*, July 31 2001, www.reap.org; A. V. Krebs, *Agribusiness Examiner*, no. 19, January 28, 1999; A. V. Krebs, It Is Plain, Cargill's Reign in the Grain Has Become Profane, *Agribusiness Examiner*, no. 9, November 12, 1998.
2. The five agribusinesses at the top of the pesticide and transgenic seed market are AstraZeneca, DuPont, Monsanto, Novartis, and Aventis. Rural Advancement Foundation International, www.rafi.org, March 16, 2000; and *Agrow*, No. 335, August 27, 1999.
3. Rural Advancement Foundation International, AgBiotech's Five Jumbo Gene Giants, www.rafi.org, March 16, 2000.
4. Quoted in J. Flint, Agricultural Industry Giants Moving Towards Genetic Monopolism, *Telepolis Magazin der Netzcultur*, June 28, 1998, www.ix.de/tp/english/inhalt/co/2385/1.html
5. US Gene "Theft" Threatens Thai rice Industry, *The Ecologist* 31:10, December 2001/January 2002.
6. Monsanto Claims Soy Gene Patent, *The Ecologist* 31:10, December 2001/January 2002; Vandana Shiva, World in a Grain of Rice, *The Ecologist* 30:9, December 2000/January 2001.
7. Leora Broydo, The Trouble with Percy, *Mother Jones*, December 13, 2000.
8. Rural Advancement Foundation International, Communiqué, March/April 1998; Ricarda A. Steinbrecher and Pat Roy Mooney, Terminator Technology: The Threat to World Food Security, *The Ecologist* 28:5, September/October 1998.
9. Brewster Kneen, *Farmaggedon: Food and the Culture of Biotechnology* (Gabriola Island, BC: New Society Publishers, 1999), p. 28.
10. Food and Agriculture Organization. *The State of Food Security in the World, 1999* (Rome: FAO, 1999).
11. UN Human Development Report, cited in UN: Gap Between Rich, Poor Grows,

New York Times, September 9, 1998.

12. Joseph Dalaker, *Poverty in the United States, 1998*, US Census Bureau, Current Population Reports, Series P60-207 (Washington, D.C.: US Government Printing Office, 1999).
13. Vandana Shiva, *Mustard or Soya? The Future of India's Edible Oil Culture* (New Delhi: Navdanya, 1998).
14. It is ironic that while many in the South are relegating whole foods to the status of poor people's food, more and more people in the industrialized world are moving in the opposite direction, often paying more for the very foods—such as whole wheat bread and brown rice—that are being abandoned in other parts of the world.
15. This is true of nonstaple foods as well. For example, in Indonesia, a country that produces some of the world's best coffee, most coffee-drinkers drink Nescafé, because it is perceived as being modern, while the vast majority of coffee produced in that country is exported to the North, where people will pay the equivalent of an Indonesian's average daily wage for a single cup.
16. Food and Agriculture Organization, *State of the World's Plant Genetic Resources* (Rome: FAO, 1996).
17. Jack Doyle, *Altered Harvest* (New York: Viking, 1985).
18. David Pimentel et al., *Environmental and Economic Benefits of Biodiversity*, unpublished manuscript, April 12, 1996, p. 3, cited in Hope Shand, *Human Nature: Agricultural Biodiversity and Farm-Based Food Security* (Ottawa, Canada: Rural Advancement Foundation International, 1997).
19. Himanshu Thakkar, Performance of Large Dams in India: The Case of Irrigation and Flood Control, in *Large Dams and Their Alternatives*, a dossier from the South Asia Consultation, Colombo, Sri Lanka, December 10–11, 1998, organized by the World Commission on Dams (Delhi: South Asia Network on Dams, Rivers & People, 1998).
20. Peter Bunyard, A Hungrier World, *The Ecologist* 29:2, March/April, 1999.
21. Peter Bunyard, How Global Warming Could Cause Northern Europe to Freeze, *The Ecologist* 29:2, March/April 1999.

a. Information taken from the Cargill website: www.cargill.com, May 15, 2000.
b. Adapted from A. V. Krebs, *Agribusiness Examiner*, no. 57, November 23, 1999.

Chapter 8: Shifting Direction

Epigraph: Michael H. Shuman, *Going Local: Creating Self-Reliant Communities in a Global Age* (New York: The Free Press, 1998), p. 6.

1. Jules Pretty, *The Living Land: Agriculture, Food and Community Regeneration in Rural Europe* (London: Earthscan, 1998).
2. Jules Pretty, *Regenerating Agriculture: Policies and Practice for Sustainability and Self-Reliance* (London: Earthscan, 1995).

3. Jules Pretty et al., *An Assessment of the External Costs of UK Agriculture* (London: Agricultural Systems, 1999).
4. McDonald's Corporation press release, McDonald's Reports Record Global Results, January 26, 2000.

a. Peter Rosset, Alternative Agriculture Works: The Case of Cuba, *Monthly Review* 50:3, July/August 1998.
b. Peter Rosset and Shea Cunningham, The Greening of Cuba, *Earth Island Journal* 10:1, Winter 1994–5, p. 23.
c. Hugh Warwick, Cuba's Organic Revolution, *The Ecologist* 29:8, December 1999.
d. Catherine Murphy, Cultivating Havana: Urban Agriculture and Food Security in the Years of Crisis," *Development Report No. 12* (Oakland, Calif.: Institute for Food and Development Policy, 1999), p. 15.
e. Murphy, *Development Report No. 12*, pp. 27–28.
f. Murphy, *Development Report No. 12*, p. 9.

Index